Sources and Studies
in the History of Mathematics and
Physical Sciences

Sources and Studies in the
History of Mathematics and Physical Sciences

K. Andersen
Brook Taylor's Work on Linear Perspective

K. Andersen
The Geometry of an Art: The History of the Mathematical Theory of Perspective from Alberti to Monge

H.J.M. Bos
Redefining Geometrical Exactness: Descartes' Transformation of the Early Modern Concept of Construction

J. Cannon/S. Dostrovsky
The Evolution of Dynamics: Vibration Theory from 1687 to 1742

B. Chandler/W. Magnus
The History of Combinatorial Group Theory

A.I. Dale
A History of Inverse Probability: From Thomas Bayes to Karl Pearson, Second Edition

A.I. Dale
Most Honourable Remembrance: The Life and Work of Thomas Bayes

A.I. Dale
Pierre-Simon Laplace, Philosophical Essay on Probabilities, Translated from the fifth French edition of 1825, with Notes by the Translator

P. Damerow/G. Freudenthal/P. McLaughlin/J. Renn
Exploring the Limits of Preclassical Mechanics

P. Damerow/G. Freudenthal/P. McLaughlin/J. Renn
Exploring the Limits of Preclassical Mechanics: A Study of Conceptual Development in Early Modern Science: Free Fall and Compounded Motion in the Work of Descartes, Galileo, and Beeckman, Second Edition

P.J. Federico
Descartes on Polyhedra: A Study of the *De Solidorum Elementis*

J. Friberg
A Remarkable Collection of Babylonian Mathematical Texts

B.R. Goldstein
The Astronomy of Levi ben Gerson (1288–1344)

H.H. Goldstine
A History of Numerical Analysis from the 16th Through the 19th Century

H.H. Goldstine
A History of the Calculus of Variations from the 17th Through the 19th Century

G. Graßhoff
The History of Ptolemy's Star Catalogue

(Continued after Index)

Anders Hald

A History of Parametric Statistical Inference from Bernoulli to Fisher, 1713-1935

 Springer

Anders Hald
University of Copenhagen
Virum 2830
Denmark

Sources and Studies Editor:
Jesper Lützen
University of Copenhagen
Department of Mathematics
Universitetsparken 5
DK-2100 Copenhagen
Denmark

Mathematics Subject Classification 2000 01A05 60-03 62-03

Library of Congress Control Number: 2006933417

ISBN-10: 0-387-46408-5   eISBN-10: 0-387-46409-3
ISBN-13: 978-0387-46408-4   eISBN-13: 978-0387-46409-1

Printed on acid-free paper.

9 8 7 6 5 4 3 2 1

springer.com

Contents

Preface XIV

1   The Three Revolutions in Parametric Statistical Inference .   1
    1.1   Introduction   1
    1.2   Laplace on Direct Probability, 1776–1799   1
    1.3   The First Revolution: Laplace 1774–1786   2
    1.4   The Second Revolution: Gauss and Laplace 1809–1828   4
    1.5   The Third Revolution: R.A. Fisher 1912–1956   6

Part I BINOMIAL STATISTICAL INFERENCE
The Three Pioneers: Bernoulli (1713), de Moivre (1733),
and Bayes (1764)

2   James Bernoulli's Law of Large Numbers for the
    Binomial, 1713, and Its Generalization   11
    2.1   Bernoulli's Law of Large Numbers for the Binomial, 1713   11
    2.2   Remarks on Further Developments   14

3   De Moivre's Normal Approximation to the Binomial,
    1733, and Its Generalization   17
    3.1   De Moivre's Normal Approximation to the Binomial, 1733   17
    3.2   Lagrange's Multivariate Normal Approximation to the
          Multinomial and His Confidence Interval for the Binomial
          Parameter, 1776   22
    3.3   De Morgan's Continuity Correction, 1838   24

4   Bayes's Posterior Distribution of the Binomial Parameter
    and His Rule for Inductive Inference, 1764   25
    4.1   The Posterior Distribution of the Binomial Parameter, 1764 . .   25
    4.2   Bayes's Rule for Inductive Inference, 1764   27

Part II STATISTICAL INFERENCE BY INVERSE PROBABIL-ITY. Inverse Probability from Laplace (1774), and Gauss (1809) to Edgeworth (1909)

5   Laplace's Theory of Inverse Probability, 1774–1786 33
   5.1   Biography of Laplace 33
   5.2   The Principle of Inverse Probability and the Symmetry of
       Direct and Inverse Probability, 1774 35
   5.3   Posterior Consistency and Asymptotic Normality in the
       Binomial Case, 1774 38
   5.4   The Predictive Distribution, 1774–1786.................... 40
   5.5   A Statistical Model and a Method of Estimation. The Double
       Exponential Distribution, 1774 41
   5.6   The Asymptotic Normality of Posterior Distributions, 1785 ... 44

6   A Nonprobabilistic Interlude: The Fitting of Equations to Data, 1750–1805 47
   6.1   The Measurement Error Model 47
   6.2   The Method of Averages by Mayer, 1750, and Laplace, 1788 .. 48
   6.3   The Method of Least Absolute Deviations by Boscovich,
       1757, and Laplace, 1799 50
   6.4   The Method of Least Squares by Legendre, 1805 52

7   Gauss's Derivation of the Normal Distribution and the Method of Least Squares, 1809 55
   7.1   Biography of Gauss 55
   7.2   Gauss's Derivation of the Normal Distribution, 1809 57
   7.3   Gauss's First Proof of the Method of Least Squares, 1809..... 58
   7.4   Laplace's Large-Sample Justification of the Method of Least
       Squares, 1810 60

8   Credibility and Confidence Intervals by Laplace and Gauss 63
   8.1   Large-Sample Credibility and Confidence Intervals for the
       Binomial Parameter by Laplace, 1785 and 1812 63
   8.2   Laplace's General Method for Constructing Large-Sample
       Credibility and Confidence Intervals, 1785 and 1812.......... 64
   8.3   Credibility Intervals for the Parameters of the Linear Normal
       Model by Gauss, 1809 and 1816 64
   8.4   Gauss's Rule for Transformation of Estimates and Its
       Implication for the Principle of Inverse Probability, 1816 65
   8.5   Gauss's Shortest Confidence Interval for the Standard
       Deviation of the Normal Distribution, 1816 66

9   The Multivariate Posterior Distribution 67
  9.1 Bienaymé's Distribution of a Linear Combination of the
       Variables, 1838 67
  9.2 Pearson and Filon's Derivation of the Multivariate Posterior
       Distribution, 1898 68

**10   Edgeworth's Genuine Inverse Method and the Equivalence
      of Inverse and Direct Probability in Large Samples, 1908
      and 1909** 69
  10.1 Biography of Edgeworth 69
  10.2 The Derivation of the t Distribution by Lüroth, 1876, and
       Edgeworth, 1883 69
  10.3 Edgeworth's Genuine Inverse Method, 1908 and 1909 71

11   Criticisms of Inverse Probability 73
  11.1 Laplace 73
  11.2 Poisson 75
  11.3 Cournot 76
  11.4 Ellis, Boole, and Venn 77
  11.5 Bing and von Kries 79
  11.6 Edgeworth and Fisher 80

**Part III THE CENTRAL LIMIT THEOREM AND LINEAR
MINIMUM VARIANCE ESTIMATION BY LAPLACE AND
GAUSS**

**12   Laplace's Central Limit Theorem and Linear Minimum
      Variance Estimation** 83
  12.1 The Central Limit Theorem, 1810 and 1812 83
  12.2 Linear Minimum Variance Estimation, 1811 and 1812 85
  12.3 Asymptotic Relative Efficiency of Estimates, 1818 88
  12.4 Generalizations of the Central Limit Theorem 90

13   Gauss's Theory of Linear Minimum Variance Estimation .. 93
  13.1 The General Theory, 1823 93
  13.2 Estimation Under Linear Constraints, 1828 96
  13.3 A Review of Justifications for the Method of Least Squares ... 97
  13.4 The State of Estimation Theory About 1830 98

**Part IV ERROR THEORY. SKEW DISTRIBUTIONS.
CORRELATION. SAMPLING DISTRIBUTIONS**

14 The Development of a Frequentist Error Theory 105
   14.1 The Transition from Inverse to Frequentist Error Theory 105
   14.2 Hagen's Hypothesis of Elementary Errors and His Maximum
       Likelihood Argument, 1837 106
   14.3 Frequentist Theory, Chauvenet 1863, and Merriman 1884 108

15 Skew Distributions and the Method of Moments 111
   15.1 The Need for Skew Distributions 111
   15.2 Series Expansions of Frequency Functions. The A and B Series 112
   15.3 Biography of Karl Pearson 117
   15.4 Pearson's Four-Parameter System of Continuous
       Distributions, 1895 120
   15.5 Pearson's χ^2 Test for Goodness of Fit, 1900 123
   15.6 The Asymptotic Distribution of the Moments by Sheppard,
       1899 125
   15.7 Kapteyn's Derivation of Skew Distributions, 1903 126

16 Normal Correlation and Regression 131
   16.1 Some Early Cases of Normal Correlation and Regression 131
   16.2 Galton's Empirical Investigations of Regression and
       Correlation, 1869–1890 134
   16.3 The Mathematization of Galton's Ideas by Edgeworth,
       Pearson, and Yule 141
   16.4 Orthogonal Regression. The Orthogonalization of the Linear
       Model 146

17 Sampling Distributions Under Normality, 1876–1908 149
   17.1 The Distribution of the Arithmetic Mean 149
   17.2 The Distribution of the Variance and the Mean Deviation by
       Helmert, 1876 149
   17.3 Pizzetti's Orthonormal Decomposition of the Sum of Squared
       Errors in the Linear-Normal Model, 1892 153
   17.4 Student's t Distribution by Gosset, 1908 154

Part V THE FISHERIAN REVOLUTION, 1912–1935

18 Fisher's Early Papers, 1912–1921 159
   18.1 Biography of Fisher 159
   18.2 Fisher's "Absolute Criterion," 1912 163
   18.3 The Distribution of the Correlation Coefficient, 1915, Its
       Transform, 1921, with Remarks on Later Results on Partial
       and Multiple Correlation 165
   18.4 The Sufficiency of the Sample Variance, 1920 172

19 The Revolutionary Paper, 1922 175
  19.1 The Parametric Model and Criteria of Estimation, 1922 175
  19.2 Properties of the Maximum Likelihood Estimate 178
  19.3 The Two-Stage Maximum Likelihood Method and Unbiasedness 182

**20 Studentization, the F Distribution, and the Analysis of
    Variance, 1922–1925** 185
  20.1 Studentization and Applications of the t Distribution 185
  20.2 The F Distribution 187
  20.3 The Analysis of Variance 188

**21 The Likelihood Function, Ancillarity, and Conditional
    Inference** 193
  21.1 The Amount of Information, 1925 193
  21.2 Ancillarity and Conditional Inference 194
  21.3 The Exponential Family of Distributions, 1934 194
  21.4 The Likelihood Function 195

References 199

Subject Index 217

Author Index 221

Preface

This is an attempt to write a history of parametric statistical inference. It may be used as the basis for a course in this important topic. It should be easy to read for anybody having taken an elementary course in probability and statistics.

The reader wanting more details, more proofs, more references, and more information on related topics may find so in my previous two books: *A History of Probability and Statistics and Their Applications before 1750,* Wiley, 1990, and *A History of Mathematical Statistics from 1750 to 1930,* Wiley, 1998.

The text contains a republication of pages 488–489, 494–496, 612–618, 620–626, 633–636, 652–655, 670–685, 713–720, and 734–738 from A. Hald: *A History of Mathematical Statistics from 1750 to 1930,* Copyright © 1998 by John Wiley & Sons, Inc. This material is used by permission of John Wiley & Sons, Inc.

I thank my granddaughter Nina Hald for typing the first version of the manuscript. I thank Professor Michael Sørensen, Department of Applied Mathematics and Statistics, University of Copenhagen for including my book in the Department's series of publications. My sincere thanks are due to Søren Johansen, Professor of mathemtical statistic at the University of Copenhagen for discussions of the manuscript leading to many improvements and furthermore for detailed help with the editing of the book.

October 2006

Anders Hald
Bagsværdvej 248, 3.-311
2800 Bagsværd
Denmark

James Bernoulli (1654–1705)   Abraham de Moivre (1667–1754)

P. S. LAPLACE.

Pierre Simon Laplace (1749–1827)

Carl Frederich Gauss (1777–1855)   Ronald Aylmer Fisher (1890–1962)

1

The Three Revolutions in Parametric Statistical Inference

1.1 Introduction

The three revolutions in parametric statistical inference are due to Laplace [148], Gauss and Laplace (1809–1811), and Fisher [67].

We use $p(\cdot)$ generically to denote a frequency function, continuous or discontinuous, and $p(x|\theta)$ to denote a statistical model defined on a given sample space and parameter space. Let $\underline{x} = (x_1, \dots, x_n)$ denote a sample of n independent observations. From the model we can find the sampling distribution of the statistic $t(\underline{x})$, and from $p(t|\theta)$ we can find probability limits for t for any given value of θ. This is a problem in direct probability, as it was called in the nineteenth century.

In inverse probability the problem is to find probability limits for θ for a given value of $\underline{x}$. Bayes [3] was the first to realize that a solution is possible only if θ itself is a random variable with a probability density $p(\theta)$. We can then find the conditional distributions $p(\theta|\underline{x})$ and $p(\theta|t)$, which can be used to find probability limits for θ for any given value of $\underline{x}$. Independently of Bayes, Laplace [148] gave the first general theory of statistical inference based on inverse probability.

1.2 Laplace on Direct Probability, 1776–1799

At the same time as he worked on inverse probability Laplace also developed methods of statistical inference based on direct probability. At the time the problems in applied statistics were mainly from demography (rates of mortality and the frequency of male births) and from the natural sciences (distribution of errors and laws of nature). It was therefore natural for Laplace to create a theory of testing and estimation comprising relative frequencies, the arithmetic mean, and the linear model, which we write in the form $y = X\beta + \varepsilon$, where $y = [y_1, \dots, y_n]'$ denotes the vector of observations, $\beta = [\beta_1, \dots, \beta_m]'$

the unknown parameters, $\varepsilon = [\varepsilon_1, \ldots, \varepsilon_n]'$ the independently distributed errors, and $X = [x_1, \ldots, x_m]$ the m column vectors of the matrix of coefficients, which are supposed to be given numbers. We also write $y = Xb + e$, where b is an estimate of β and e denotes the corresponding residuals.

The error distributions discussed at the time were symmetric with known scale parameter, the most important being the rectangular, triangular, quadratic, cosine, semi-circular, and the double exponential. The normal distribution was not yet invented.

The arithmetic mean was ordinarily used as an estimate of the location parameter. Laplace [149] solved the problem of finding the distribution of the mean by means of the convolution formula. However, this was only a solution in principle because all the known error distributions, apart from the rectangular, led to unmanageable distributions of the mean. He also gave the first test of significance of a mean based on the probability of a deviation from the expected value as large or larger than the observed, assuming that the observations are rectangularly distributed.

Three methods of fitting the linear model to data without specification of the error distribution were developed: the method of averages by Mayer [174] and Laplace [152] requiring that $\sum w_{ik} e_i = 0$, where the ws are suitably chosen weights and the number of equations equals the number of unknown parameters; the method of least absolute deviations by Boscovich [17] and Laplace [151], were $\sum w_i e = 0$ and $\sum w_i |e_i|$ is minimized for the two-parameter model; and the method of minimizing the largest absolute deviation by Laplace [151], that is, $\min_\beta \max_i |y_i - \beta x_i|$. He evaluated the results of such analyses by studying the distribution of the residuals.

1.3 The First Revolution: Laplace 1774–1786

Turning to inverse probability let us first consider two values of the parameter and the corresponding direct probabilities. Laplace's principle says, that if $\underline{x}$ is more probable under θ_2 than under θ_1 and $\underline{x}$ has been observed, then the probability of θ_2 being the underlying value of θ (the cause of $\underline{x}$) is larger than the probability of θ_1. Specifically, Laplace's principle of inverse probability says that

$$\frac{p(\theta_2|\underline{x})}{p(\theta_1|\underline{x})} = \frac{p(\underline{x}|\theta_2)}{p(\underline{x}|\theta_1)}$$

for all (θ_1, θ_2), or equivalently that

$$p(\theta|\underline{x}) \propto p(\underline{x}|\theta);$$

that is, inverse probability is proportional to direct probability. In the first instance Laplace formulated this principle intuitively; later he proved it under the supposition that the prior density is uniform on the parameter space. Fisher [67] introduced the likelihood function $L_x(\theta)$, defined as proportional to $p(\theta|x)$, to avoid the theory of inverse probability. The relation between the theories of Laplace and Fisher is illustrated in the following diagram

$$
\begin{array}{ccccc}
p(\theta|\underline{x}) & \propto & p(\underline{x}|\theta) & \propto & L_x(\theta) \\
Inverse\ probability & & Direct\ probability & & Likelihood \\
& Laplace & | & Fisher &
\end{array}
$$

The history of statistical inference is about $p(\underline{x}|\theta)$ and its two interpretations, or in modern terminology about sampling distributions, posterior distributions, and the likelihood function. The mathematical parts of the three topics are closely related and a new result in any of the three fields has repercussions in the other two.

Based on Laplace's principle it is a matter of mathematical technique to develop a theory of testing, estimation, and prediction, given the model and the observations. Laplace did so between 1774 and 1786. To implement the theory for large samples, Laplace developed approximations by means of asymptotic expansion of integrals, both for tail probabilities and for probability integrals over an interval containing the mode. Using the Taylor expansion about the mode $\hat{\theta}$, and setting $\log L_x(\theta) = l(\theta)$, he found

$$
\log p(x|\theta) = constant + l(\theta) = constant + l(\hat{\theta}) + \frac{1}{2}(\theta - \hat{\theta})^2 l''(\hat{\theta}) + \cdots ,
$$

which shows that θ is asymptotically normal with mean $\hat{\theta}$ and variance $[-l''(\hat{\theta})]^{-1}$.

In this way Laplace proved for the binomial distribution that the most probable value of θ equals the observed relative frequency h and that θ is asymptotically normal with mean h and variance $h(1 - h)/n$. Moreover, to test the significance of the difference $h_1 - h_2$ between two relative frequencies, he showed that $\theta_1 - \theta_2$ is asymptotically normal with mean $h_1 - h_2$ and variance $h_1(1 - h_1)/n_1 + h_2(1 - h_2)/n_2$, which led him to the large sample test of significance used today.

There is, however, an inconsistency in Laplace's theory of estimation. For the binomial and the multinomial distributions he uses the most probable value as estimate, but in the measurement error model he introduces a new criterion to estimate the location parameter, namely to minimize the posterior expected loss, using the absolute deviation as loss function. He proves that this leads to the posterior median as estimator. His justification for this procedure is that the absolute deviation is the natural measure of the goodness of the estimate and that it corresponds to the gambler's expected loss in a game of chance.

The introduction of a loss function proved to be a serious mistake, which came to hamper the development of an objective theory of statistical inference to the present day. It is of course the beginning of the split between inference and decision theory.

To try out the new method Laplace chose the simplest possible error distribution with infinite support, the double exponential distribution. For three observations he found that the estimate is a root of a polynomial equation of the fifteenth degree. It must have been a great disappointment for him that the combination of the simplest possible error distribution and the simplest possible loss function led to an unmanageable solution, even for three observations.

In 1799, at the end of the first revolution, one important problem was still unsolved: the problem of the arithmetic mean. Applying all the known methods of estimation to all the known error distributions led to estimates of the location parameter different from the mean. Nevertheless, in practice everybody used the mean.

1.4 The Second Revolution: Gauss and Laplace 1809–1828

The second revolution began in 1809–1810 with the solution of the problem of the mean, which gave us two of the most important tools in statistics, the normal distribution as a distribution of observations, and the normal distribution as an approximation to the distribution of the mean in large samples.

In 1809 Gauss asked the question: Does there exist an error distribution leading to the mean as estimate of the location parameter according to the principle of inverse probability? Gauss did not make the mistake of Laplace of introducing a loss function; instead he used the most probable value of the parameter as estimate. Setting the posterior mode equal to the arithmetic mean of the observations he got a functional equation with the normal distribution as solution. The normal distribution thus emerged as a mathematical construct, and Gauss did not compare the new error distribution with observations.

Assuming that the observations are normally distributed he found that the most probable value of the location parameter is obtained by minimizing the exponent $\sum (y_i - \theta)^2$, which naturally leads to the mean. If θ is a linear function of m parameters, $\theta = X\beta$, the estimates are found by minimizing the sum of the squared errors $(Y - X\beta)'(Y - X\beta)$. Assuming the variance of the ys to be known, Gauss solved the estimation problems for the linear-normal model and derived the multivariate normal distribution of the parameters.

Before having seen Gauss's book, Laplace [155] published a paper in which he derived the first version of the central limit theorem, which says that regardless of the shape of the error distribution, if only the variance is finite, the mean will be approximately normally distributed in large samples. As his immediate reaction to Gauss's results Laplace made two remarks [156]:

(1)   If the error distribution is normal, then the posterior distribution is normal and the posterior mean and median are equal. Hence, the method of least squares follows from my method of estimation as a special case.

(2)   If the error distribution has finite variance, but is otherwise unknown, then the central limit theorem gives a large-sample justification for the method.

Hence, in the first instance, both Gauss and Laplace used inverse probability in their derivations of the method of least squares.

But already in 1811 Laplace gave an alternative derivation based on direct probability using the asymptotic normality of a linear combination of observations and minimizing the expected absolute error, which for the normal distribution is proportional to the expected squared error.

In 1823 and 1828 Gauss supplemented Laplace's large-sample frequentist theory by a small-sample theory. Like Laplace he replaced the assumption of normality with the weaker assumption of finite variance, but in contradistinction to Laplace he used squared error as loss function because of its greater mathematical simplicity. He then developed the theory of linear, unbiased, minimum variance estimation for the linear model in the form known today.

Hence, they both gave up the normality assumption as too restrictive.

Gauss's two proofs both became popular and existed beside each other in spite of their contradictory assumptions. One reason for this may be the following argument due to Laplace.

In 1812 Laplace [159] made an important observation on the equivalence of direct and inverse probability for finding large-sample limits for the binomial parameter. Direct probability leads to the limits for the relative frequency h of the form

$$h \sim \theta \pm \sqrt{\theta(1 - \theta)/n},$$

disregarding terms of the order of $1/n$. But for this order of approximation the limits may also be written as

$$h \sim \theta \pm \sqrt{h(1 - h)/n},$$

which solved for θ gives

$$\theta \sim h \pm \sqrt{h(1 - h)/n}.$$

However, these limits are the same as those following from inverse probability. Generalizing this argument, the probability limits for the estimate t become

$$t \sim \theta \pm \sigma\sqrt{n},$$

and for the estimate s

$$s \sim \sigma \pm \kappa/\sqrt{n}.$$

Combining these relations we get

$$t \sim \theta \pm s/\sqrt{n}$$

which leads to the limits for θ,

$$\theta \sim t \pm s/\sqrt{n}.$$

This kind of reasoning explains why the methods of direct and inverse probability could coexist in statistical practice without serious conflict for about a hundred years.

For large samples the normal distribution could be used to find probability or confidence limits. For moderately large samples the so-called 3σ-limits became a standard procedure in estimation and testing as a safeguard against deviations from normality.

During the following period the application of statistical methods was extended to the social and biological sciences in which variation among individuals, instead of errors, was studied by means of skew frequency curves, and the measurement error model was replaced by linear regression and correlation.

Two systems of frequency curves were developed: Pearson's system of skew frequency curves, and Kapteyn's system of transformations to obtain normality.

Correspondingly, a new method of estimation was developed which may be called the analogy method. Pearson equated the empirical moments to the theoretical moments and thus got as many nonlinear equations as parameters to be estimated. Kapteyn equated the empirical and theoretical percentiles.

1.5 The Third Revolution: R.A. Fisher 1912–1956

At the beginning of the present century the theory of statistical inference thus consisted of a large number of ad hoc methods, some of them contradictory, and the small-sample theory was only in a rudimentary state. Some important questions were as follows.

How to choose between direct and inverse probability methods?

How to choose between various loss functions?

How to choose between various statistics for use in the analogy method?

How to find probability limits for the parameters from direct probability methods?

These problems were attacked and most of them solved by Fisher between 1922 and 1936.

He turned the estimation problem upside down by beginning with requirements to estimators. He formulated the criteria of consistency, efficiency, and sufficiency, the last concept being new.

Having thus defined the properties of good estimators he turned to a criticism of the existing methods of estimation. He showed that the inverse probability estimate depends on the parameterization of the model, which means that the resulting estimate is arbitrary. For a time this argument led to less interest in inverse probability methods.

He rejected the use of loss functions as extraneous to statistical inference.

Turning to analogy methods he showed that the method of moments in general is inefficient.

Given the model and the observations, he noted that all information on the parameters is contained in the likelihood function, and he proved the asymptotic optimality of the estimates derived from this function, the maximum likelihood estimates. Basing his inference theory on the likelihood function he avoided the arbitrariness introduced by Laplace and Gauss due to loss functions and the assumption of finite variance.

Assuming normality, he derived the t, χ^2, and F distributions, and showed how to use them in testing and interval estimation, thus solving the small-sample problems for the linear-normal model.

He also derived the distribution of the correlation coefficient and the partial correlation coefficients in normal samples.

He initiated the theory of ancillary statistics and conditional inference. Large-sample probability limits for a parameter were found by what today is called a pivotal statistic. By an ingenious use of the pivotal argument, Fisher derived what he called fiducial limits for a parameter, for example, by means of the t distribution. Fisher explained the new statistical ideas and techniques in an aggressive and persuasive language, which led to acceptance of his theories within a rather short period of time, not alone among mathematical statisticians, but also among research workers in general. A large part of mathematical statistics since 1922 has consisted in an elaboration of Fisher's ideas, both in theory and practice.

Because of the fundamental relation between the posterior density and the likelihood function many of Fisher's asymptotic results are identical to those of Laplace from a mathematical point of view, only a new interpretation is required. Fisher never acknowledged his debt to Laplace.

Figure 1.5.1 indicates how the ideas of Laplace, Gauss, and Fisher have influenced statistical theory today.

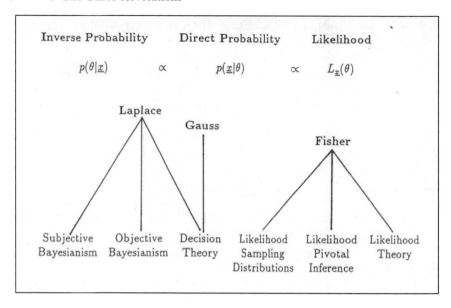

Fig. 1.5.1. The diagram indicates the relation among the ideas of Laplace, Gauss, and Fisher.

Part I

BINOMIAL STATISTICAL INFERENCE
The Three Pioneers: Bernoulli (1713), de Moivre (1733), and Bayes (1764)

James Bernoulli's Law of Large Numbers for the Binomial, 1713, and Its Generalization

2.1 Bernoulli's Law of Large Numbers for the Binomial, 1713

James Bernoulli (1654–1705) graduated in theology from the University of Basel in 1676; at the same time he studied mathematics and astronomy. For the next seven years he spent most of his time traveling as tutor and scholar in Switzerland, France, the Netherlands, England, and Germany. Returning to Basel in 1683 he lectured on mathematics and experimental physics and in 1687 he became professor of mathematics at the university. He and his younger brother John made essential contributions to Leibniz's new infinitesimal calculus. He left his great work on probability *Ars Conjectandi* (The Art of Conjecturing) unfinished; it was published in 1713.

At the times of Bernoulli the doctrine of chances, as probability theory then was called, had as its main aim the calculation of a gambler's expected gain in a game of chance. Because of the symmetry of such games, all possible outcomes were considered equally probable which led to the classical definition of probability as the ratio of the number of favorable cases to the total number of possible cases. Bernoulli's great vision was to extend the doctrine of chances to a probability theory for treating uncertain events in "civil, moral and economic affairs." He observes that in demography, meteorology, insurance, and so on it is impossible to use the classical definition, because such events depend on many causes that are hidden for us. He writes ([6], p. 224):

> But indeed, another way is open to us here by which we may obtain what is sought; and what you cannot deduce *a priori*, you can at least deduce *a posteriori* – i.e. you will be able to make a deduction from the many observed outcomes of similar events. For it may be presumed that every single thing is able to happen and not to happen in as many cases as it was previously observed to have happened or not to have happened in like circumstances.

Bernoulli refers to the well-known empirical fact that the relative frequency of an event, calculated from observations taken under the same circumstances, becomes more and more stable with an increasing number of observations. Noting that the statistical model for such observations is the binomial distribution, Bernoulli asks the fundamental question: Does the relative frequency derived from the binomial have the same property as the empirical relative frequency? He proves that this is so and concludes that we may then extend the application of probability theory from games of chance to other fields where stable relative frequencies exist. We give Bernoulli's theorem and proof in modern formulation.

Consider n independent trials each with probability p for "success," today called Bernoulli trials. The number of successes, s_n say, is binomially distributed $(n, p), 0 < p < 1$. Assuming that np and $n\varepsilon$ are positive integers, Bernoulli proves that the relative frequency $h_n = s_n/n$ satisfies the relation

$$P_n = P\{|h_n - p| \leq \varepsilon\} > c/(c+1) \text{ for any } c > 0, \tag{2.1}$$

if

$$n \geq \frac{m(1+\varepsilon) - q}{(p+\varepsilon)\varepsilon} \vee \frac{m(1+\varepsilon) - p}{(q+\varepsilon)\varepsilon}, \tag{2.2}$$

where m is the smallest integer satisfying the inequality

$$m \geq \frac{\log(c(q-\varepsilon)/\varepsilon)}{\log((p+\varepsilon)/p)} \vee \frac{\log(c(p-\varepsilon)/\varepsilon)}{\log((q+\varepsilon)/q)}. \tag{2.3}$$

Hence, for any fixed value of p and ε, however small, and for c tending to infinity, the lower bound for n tends to infinity and P_n tends to 1. This is the law of large numbers for binomially distributed variables; h_n tends in probability to p.

However, in addition to the limit theorem Bernoulli provides a lower bound for n. As an example he takes $p = 0.6, \varepsilon = 0.02$, and $c = 1000$ which leads to

$$P(0.58 \leq h_n \leq 0.62) > 1000/1001 \text{ for } n \geq 25{,}550.$$

In the proof Bernoulli sets

$$s_n - np = x, \ x = -np, -np + 1, \ldots, nq, \tag{2.4}$$

and $n\varepsilon = k, \ k = 1, 2, \ldots$, so that $P_n = P\{|x| \leq k\}$. The distribution of x is

$$f(x) = \binom{n}{np + x} p^{np+x} q^{nq-x}.$$

The inequality $P_n > c/(c+1)$ is replaced by the equivalent $P_n/(1 - P_n) > c$, which means that the central part of the binomial should be larger than c times the sum of the tails. Disregarding $f(0)$, it is thus sufficient to require that this inequality hold for each tail; that is,

$$\sum_{l}^{k} f(x) \geq c \sum_{k+1}^{nq} f(x) \qquad (2.5)$$

for the right tail. The result for the left tail is obtained by interchanging p and q.

Bernoulli investigates the properties of the binomial by means of the ratio

$$\frac{f(x)}{f(x+1)} = \frac{np+x+1}{nq-x} \frac{q}{p}, \quad x = 0, 1, \ldots, nq-1, \qquad (2.6)$$

which is an increasing function of x. It follows that $f(x)$ is decreasing for $x \geq 0$ and that

$$f(0)/f(k) < f(x)/f(x+k), \quad x \geq 1. \qquad (2.7)$$

Bernoulli uses the crude upper bound

$$\sum_{k+1}^{nq} f(x) \leq \frac{nq-k}{k} \sum_{k+1}^{2k} f(x), \qquad (2.8)$$

so that (2.5) is replaced by

$$\sum_{1}^{k} f(x) \geq c \frac{nq-k}{k} \sum_{k+1}^{2k} f(x), \qquad (2.9)$$

which by means of (2.7) leads to

$$\frac{f(0)}{f(k)} \geq c \frac{nq-k}{k}. \qquad (2.10)$$

Hence, the problem of evaluating the ratio of two sums has been reduced to the evaluation of the ratio of two binomial probabilities.

From (2.6) it follows that

$$\frac{f(k)}{f(0)} = \prod_{i=1}^{k} (1 + \frac{k+1-i}{np})/(1 - \frac{k-i}{nq}). \qquad (2.11)$$

The k factors are decreasing with i and lie between $\{1 + k/np)\}/\{1 - (k/nq)\}$ and 1. To get a closer bound for $f(0)/f(k)$, Bernoulli chooses n so large that there is an m for which

$$f(0)/f(k) \geq [1 + (k/np)]^m 1^{k-m}.$$

The problem is thus reduced to solving the inequality

$$[1 + (k/np)]^m \geq c(nq-k)/k$$

with respect to m, and solving the equation

$$1 + \frac{k+1-m}{kp}\varepsilon = \left(1 + \frac{\varepsilon}{p}\right)\left(1 - \frac{k-m}{kq}\varepsilon\right)$$

with respect to $k = n\varepsilon$. The solution is given by (2.2) and (2.3).

James Bernoulli's ideas and his proof of the law of large numbers became a great inspiration for probabilists and statisticians for the next hundred years.

2.2 Remarks on Further Developments

Bernoulli's lower bound for n is rather large because of two crude approximations. First, he requires that the basic inequality hold for each tail separately instead of for the sum only; see (2.5). Second, he uses an arithmetic approximation for the tail probability instead of a geometric one; see (2.8). These defects were corrected by Chebyshev [29] and Nicolas Bernoulli [6], respectively. The law of large numbers may be considered as a corollary to Laplace's central limit theorem, which holds for sums of random variables, discrete or continuous. It was generalized by Poisson [222] to sums of random variables with different distributions so the sample mean $\bar{x}_n$ is asymptotically normal with mean $\bar{\mu}_n$ and $V(\bar{x}_n) = \sum \sigma_i^2/n^2$, which is supposed to be of order n^{-1}. Hence,

$$P\{|\bar{x}_n - \bar{\mu}_n| < \varepsilon\} \cong \Phi(u) - \Phi(-u), \ u = \varepsilon/\sqrt{V(\bar{x}_n)}, \qquad (2.12)$$

which tends to 1 as $n \to \infty$. As a special case Poisson considers a series of trials with varying probabilities of success, $p_1, p_2, \ldots$, today called Poisson trials. It follows that the relative frequency of successes in n trials, h_n say, tends in probability to $(p_1 + \cdots + p_n)/n$ provided $\sum p_i q_i \to \infty$ as $n \to \infty$, which is the case if the ps are bounded away from 0 and 1. It is supposed that $\bar{p}_n$, the average of the ps, tends to a constant. It was Poisson who introduced the name "the law of large numbers" for the fact that $|\bar{x}_n - \bar{\mu}_n|$ converges in probability to zero.

Chebyshev [29] proves the law of large numbers for Poisson trials by a generalization of Bernoulli's proof. He finds that

$$P_n = P(|h_n - \bar{\mu}_n| < \varepsilon) < 1 - \delta_n,$$

δ_n being a function of n, $\bar{p}_n$, and ε, which tends exponentially to zero as $n \to \infty$. He determines a lower bound for n such that $P_n > c/(c+1)$, setting each tail probability smaller than $1/[2(c+1)]$. For the right tail he finds

$$n > \ln\left[\frac{q(c+1)}{\varepsilon}\sqrt{\frac{p+\varepsilon}{q-\varepsilon}}\right] \Big/ \ln\left[\left(\frac{p+\varepsilon}{p}\right)^{p+\varepsilon}\left(\frac{q-\varepsilon}{q}\right)^{q-\varepsilon}\right], \ p = \bar{p}_n. \quad (2.13)$$

The lower bound for the left tail is found by interchanging p and q. Chebyshev's lower bound is approximately equal to $2pq$ times Bernoulli's bound; for Bernoulli's example, (2.13) gives $n > 12{,}243$ compared with Bernoulli's 25,550.

Independently, Bienaymé [14] and Chebyshev [31] prove the law of large numbers without recourse to the central limit theorem. For the random variable x with mean μ and standard deviation σ, $0 < \sigma < \infty$, they prove the inequality

$$P\{|x - \mu| \leq t\sigma\} \geq 1 - t^{-2}, \text{ for any } t > 0. \tag{2.14}$$

Hence,

$$P\left(|\overline{x}_n - \overline{\mu}_n|\right) \leq t\sqrt{V(\overline{x}_n)} \geq 1 - t^{-2}, \tag{2.15}$$

from which the law of large numbers immediately follows.

Khintchine [138] proves that $\overline{x}_n$ tends in probability to μ if the xs are independently and identically distributed with finite expectation μ. Hence, in this case the law of large numbers holds even if the variance does not exist.

3

De Moivre's Normal Approximation to the Binomial, 1733, and Its Generalization

3.1 De Moivre's Normal Approximation to the Binomial, 1733

Abraham de Moivre (1667–1754) was of a French Protestant family; from 1684 he studied mathematics in Paris. The persecution of the French Protestants caused him at the age of 21 to seek asylum in England. For the rest of his life he lived in London, earning his livelihood as a private tutor of mathematics and later also as a consultant to gamblers and insurance brokers. He became a prominent mathematician and a Fellow of the Royal Society in 1697, but he never got a university appointment as he had hoped. He wrote three outstanding books: *Miscellanea Analytica* (1730), containing papers on mathematics and probability theory; *The Doctrine of Chances: or, A Method of Calculating the Probability of Events in Play* (1718, 1738, 1756); and *Annuities upon Lives* (1725, 1743, 1750, 1752), each new edition being an enlarged version of the previous one. His *Doctrine* contained new solutions to old problems and an astounding number of new results; it was the best textbook on probability theory until Laplace [159]. Here we only discuss his two proofs of Bernoulli's law of large numbers and his two approximations to the binomial.

De Moivre ([182]; [184], Problem 87) considers a game with probability p of success in which a spectator gains $|s_n - np|$ if the outcome is s_n successes in n trials, np being an integer. He proves that the expected gain equals

$$D_n = E\left(|s_n - np|\right) = 2npq \binom{n}{np} p^{np} q^{nq} \simeq \sqrt{2npq/\pi},$$

a quantity known today as the mean deviation of the binomial. The limit is obtained by means of his (1733) normal approximation to $b(np, n, p)$. The average gain per trial is

$$D_n/n = E(|h_n - p|) \simeq \sqrt{2pq/\pi n}. \tag{3.1}$$

De Moivre then gives another interpretation of this result, namely as a measure of the dispersion of the random variable h_n around the true value p.

This is the first time that such a measure is defined and discussed. Because D_n/n is a decreasing function of n, de Moivre concludes that h_n converges in probability to p. However, he does not explain how the relation $P(|h_n - p| \le \varepsilon) \to 1$ follows from (3.1). By a similar argument as that leading to the Bienaymé-Chebyshev inequality we have $P_n > 1 - (D_n/n\varepsilon)$. De Moivre adds that a more precise proof of Bernoulli's limit theorem will be given by means of his normal approximation to the binomial.

Like the Bernoullis, de Moivre wanted an approximation to the sum

$$P_n(d) = P(|x - np| \le d) = \sum_{np-d}^{np+d} b(x,n,p), \ d = 1,2,\dots$$

for large n, but unlike them he began by approximating $b(x,n,p)$. Between 1721 and 1730 he worked hard on this problem and succeeded in deriving an asymptotic expansion for $b(x,n,p)$ as $n \to \infty$; his proofs are given in the *Miscellanea Analytica* [182]. He uses the same method of proof in his various attempts; we illustrate this method by giving his proof of Stirling's formula for $m!$, which he [182] derived independently of Stirling [253].

Taking the logarithm of

$$\frac{m^{m-1}}{(m-1)!} = \prod_{i=1}^{m-1}\left(1 - \frac{i}{m}\right)^{-1}, \ m = 2,3\dots,$$

he gets

$$\ln\frac{m^{m-1}}{(m-1)!} = \sum_{k=1}^{\infty}\frac{1}{km^k}\sum_{i=1}^{m-1}i^k \tag{3.2}$$

$$= m - \frac{1}{2}\ln m - \ln\sqrt{2\pi} - \sum_{k=1}^{\infty}\frac{B_{2k}}{(2k-1)(2k)}m^{1-2k},$$

where Bernoulli's summation formula has been used for $\sum i^k$, $\{B_{2k}\}$ are the Bernoulli numbers, and $\ln\sqrt{2\pi}$ is introduced by means of the relation

$$\ln\sqrt{2\pi} = 1 - \sum_{k=1}^{\infty}B_{2k}/(2k-1)(2k), \tag{3.3}$$

which is due to Stirling. Solving for $\ln(m-1)!$ and adding $\ln m$ de Moivre gets

$$\ln m! \sim (m + \frac{1}{2})\ln m - m + \ln\sqrt{2\pi} + \sum_{k=1}^{\infty}\frac{B_{2k}}{(2k-1)(2k)}m^{1-2k}, \tag{3.4}$$

and

$$m! \sim \sqrt{2\pi m}\ m^m \exp\left(-m + \frac{1}{12m} - \frac{1}{360m^3} + \cdots\right), \tag{3.5}$$

which today is called Stirling's formula.

The Bernoullis had shown that the evaluation of $P_n/(1 - P_n)$ depends essentially on $f(0)/f(d)$ and $f(0)/f(-d)$. De Moivre begins by studying the symmetric binomial, thus avoiding the complications due to the skewness. He notes that the properties of $b(x, n, p)$ for $n \to \infty$ follow easily from the properties of $b(x, n, \frac{1}{2})$ because

$$b(n, x, p) = b(n, x, \frac{1}{2})(2p)^x (2q)^{n-x}. \tag{3.6}$$

Let $b(m + d)$ denote the symmetric binomial for $n = 2m$, that is,

$$b(m + d) = \binom{2m}{m + d} 2^{-m}, \ |d| = 0, 1, \ldots, m, \ m = 1, 2, \ldots,$$

so that $b(m)/b(m + d)$ corresponds to $f(0)/f(d)$. It follows that

$$\ln b(m) = (-2m + 1)\ln 2 + \sum_{i=1}^{m-1} \ln \frac{1 + i/m}{1 - i/m} \tag{3.7}$$

$$= (2m - \frac{1}{2})\ln(2m - 1) - 2m\ln(2m) + \ln 2 - \frac{1}{2}\ln(2\pi) + 1 + \cdots,$$

and

$$\ln \frac{b(m)}{b(m + d)} = \ln(1 + d/m) + \sum_{i=1}^{d-1} \ln \frac{1 + i/m}{1 - i/m} \tag{3.8}$$

$$= (m - d - \frac{1}{2})\ln(m - d - 1)$$

$$+ (m - d + \frac{1}{2})\ln(m - d + 1) - 2m \ln(m) + \ln(1 + d/m) + \cdots.$$

The two series are obtained by expanding the logarithm of the individual terms in powers of i/m and using Bernoulli's formula for the summation of integers, just as in (3.2). The following terms are of the order m^{-k} and $(m \pm d)^{-k}$, $k = 1, 2, \ldots$, respectively, for $d = o(m)$ and $m \to \infty$. De Moivre [182] writes that he obtained the main terms in 1721 with the modification that he had determined the constant term in (3.7) to 0.7739 instead of the correct value 0.7742 because he at the time did not know (3.3). Combining the two series he gets an approximation to the symmetric binomial from which the skew binomial is found by means of (3.6). For large n the main term is

$$b(x, n, p) \sim \frac{n^{n+1/2}}{\sqrt{2\pi}x(x - 1)^{x-1/2}(n - x + 1)^{n-x+1/2}} p^x q^{n-x}, \ x = np+d, \ d = o(n), \tag{3.9}$$

which is easy to calculate. However, he did not succeed in getting a simple expression for $P_n(d)$ by means of this formula.

It is shown from (3.9) that the main result of de Moivre's analysis in 1721 is an approximation to $\binom{n}{x}$. It was not until 1730 that he found an approximation to $n!$

Finally he realized (1733) that he had to sacrifice the asymptotic expansions, in which he had invested so much labor, and be content with an approximation to the main term to get an expression that could be evaluated by summation (integration). Using the series expansion of $\ln(1 \pm x)$ on the terms of (3.8) he gets

$$\lim_{m \to \infty} \ln \frac{b(m+d)}{b(m)} = -\frac{d^2}{m}, \ d = O(\sqrt{m}),$$

so

$$b(m+d) \sim (\pi m)^{-1/2} \exp(-d^2/m). \tag{3.10}$$

He then obtains the desired result by approximating the sum of $b(m+d)$ by the corresponding integral.

Without proof he states the general formula

$$b(np+d, n, p) \sim (2\pi npq)^{-1/2} \exp(-d^2/2npq), \ d = O(\sqrt{n}). \tag{3.11}$$

The proof is simple. Stirling's formula gives immediately

$$f(0) = b(np, n, p) \sim (2\pi npq)^{-1/2}.$$

Using that

$$\frac{f(0)}{f(d)} = \frac{b(np)}{b(np+d)} = (1 + d/np) \prod_{i=1}^{d-1} \frac{1+i/np}{1-i/nq},$$

and

$$\ln \frac{1+i/np}{1-i/nq} = \frac{i}{npq} + \cdots,$$

it follows that

$$\ln \frac{b(np)}{b(np+d)} = \frac{d^2}{2npq} + \cdots,$$

which completes the proof.

De Moivre's result may be written as

$$\sqrt{npq} \, b(x, n, p) \sim \phi(u), \ u = (x - np)/\sqrt{npq} = O(1), \tag{3.12}$$

which shows that the limit distribution of the standardized variable u for $n \to \infty$ is the same for all binomial distributions regardless of the value of p, if only p is bounded away from 0 and 1. This is the first appearance of the normal distribution in statistics.

The problem is, however, under what conditions this property holds for finite values of n. It is no wonder that the logarithm of the symmetric binomial can be accurately approximated by a parabola for small values of n; this is illustrated by de Moivre by two examples for $n = 900$. It is also clear that this is not so for the skew binomial and one may wonder why de Moivre did not develop a correction for skewness by including one more term in his expansion. The explanation is that de Moivre was looking for an approximation to $P_n(d)$ wherefore he was interested only in $b(np - d, n, p) + b(np + d, n, p)$ for which the positive and negative errors of the two components to a large extent compensate each other, see Fig. 3.1.1.

Replacing the sum of $b(x, n, p)$ by the corresponding integral based on (3.12) de Moivre concludes that

$$P_n(d) \cong (P\,|u| \le d/\sqrt{npq}) = 2 \int_0^{d/\sqrt{npq}} \phi(u)du, \ d > 0 \qquad (3.13)$$

or

$$P_n(d) \cong \phi(t) - \phi(-t), \ t = d/\sqrt{npq}. \qquad (3.14)$$

He shows how to calculate $P(|u| \le t)$ by a series expansion for $t \le 1$ and by numerical integration for $t > 1$ and carries out the calculation for $t = 1, 2, 3$. For the symmetric case he writes that (3.13) for $n > 100$ is "tolerably accurate, which I have confirmed by trials."

De Moivre presents examples of intervals for s_n and h_n of the form $np \pm t\sqrt{npq}$ and $p \pm t\sqrt{pq/n}$, respectively, corresponding to the probabilities (3.14) for $t = 1, 2, 3$.

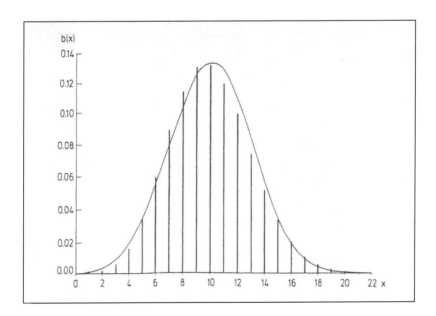

Fig. 3.1.1. The binomial distribution for $n = 100$ and $p = 0.1$ and de Moivre's normal approximation with mean 10 and standard deviation 3.

From the relation

$$P(|h_n - p| \leq \varepsilon) \sim P(|u| \leq \varepsilon\sqrt{n/pq}) \to 1, \text{ as } n \to \infty \qquad (3.15)$$

de Moivre concludes that h_n tends in probability to p.

The mathematical and numerical simplicity of de Moivre's limit theorem makes it one of the great advances in probability theory.

3.2 Lagrange's Multivariate Normal Approximation to the Multinomial and His Confidence Interval for the Binomial Parameter, 1776

The great French mathematician Joseph Louis Lagrange (1736–1813) generalizes de Moivre's result from the binomial to the multinomial distribution. Lagrange [146] considers an error distribution with k possible measurement errors, $x_1, \ldots, x_k$, occurring with probabilities $p_1, \ldots, p_k$, $\sum p_i = 1$, so that $E(x) = \sum x_i p_i = \mu$, say. He wants to estimate μ for calibrating the measuring instrument.

Let n_i be the number of times the error x_i occurs among n observations, $\sum n_i = n$, so that the sample mean is

$$\bar{x}_n = \sum x_i h_i, \ h_i = n_i/n, \ \sum h_i = 1.$$

The probability of the observed outcome is the multinomial

$$f(n_1, \ldots, n_k; p_1, \cdots, p_k) = \frac{n!}{n_1! \ldots n_k!} p_1^{n_1} \cdots p_k^{n_k},$$

which Lagrange considers as a function of the ps. Maximizing f with respect to the ps, he finds that h_i is "the most probable" value of p_i; today we would say "most likely," and that

$$f_0 = \max_{p_1, \ldots, p_k} f = f(n_1, \ldots, n_k; h_1, \ldots, h_k).$$

Setting

$$p_i = h_i + d_i/n, \ \sum d_i = 0,$$

he gets

$$f = f_0 \prod_{i=1}^{k} \left(1 + \frac{d_i}{n_i}\right)^{n_i}.$$

Assuming that $d_i = O(\sqrt{n})$ and setting $d_i = \delta_i\sqrt{n}$ he finds

$$\sum n_i \left(\ln(1 + \frac{d_i}{n_i})\right) = -\frac{1}{2}\sum \frac{\delta_i^2}{h_i} + O(n^{-1/2}).$$

Approximating the factorials by means of Stirling's formula, he obtains the large-sample approximation

$$n^{(k-1)/2} f(n_1, \ldots, n_k; p_1, \ldots, p_k) \tag{3.16}$$

$$= p(\delta_1, \ldots, \delta_k) \cong \frac{1}{(2\pi)^{(k-1)/2}(h_1 \cdots h_k)^{1/2}} \exp\left(-\frac{1}{2}\sum \frac{\delta_i^2}{h_i}\right), \quad k = 2, 3 \ldots,$$

which is a $(k-1)$-dimensional normal distribution because $\sum \delta_i = 0$.

Lagrange remarks that it is difficult to find the probability that $|\delta_i| < \rho$ for all i if $k > 2$. For $k = 2$ it follows from (3.16) that δ_1 is asymptotically normal $[0, h_1(1 - h_1)]$ so that p_1 satisfies the inequality

$$h_1 - t\sqrt{h_1(1 - h_1)/n} < p_1 < h_1 + t\sqrt{h_1(1 - h_1)/n}, \; t > 0, \tag{3.17}$$

with probability

$$P\left(|\delta_1| < t\sqrt{h_1(1 - h_1)}\right) \cong \Phi(t) - \Phi(-t). \tag{3.18}$$

This result seems to be the first example of a non-Bayesian probability statement about a parameter.

To compare the results of de Moivre and Lagrange let us write de Moivre's relation between h and p as

$$h = p + u\sqrt{p(1 - p)/n} + o(n^{-1/2}), \tag{3.19}$$

where u is asymptotically normal (0,1). Solving for p we get Lagrange's relation

$$p = h - u\sqrt{h(1 - h)/n} + o(n^{-1/2}). \tag{3.20}$$

Hence,

$$u = (h - p)\sqrt{n}/\sqrt{h(1 - h)} \tag{3.21}$$

and solving the equation

$$P(|u| < t) = \Phi(t) - \Phi(-t)$$

with respect to p, we get (3.17) and (3.18).

This mode of reasoning is a special case of Fisher's [80] fiducial probability, which Fisher ([86] p. 51) later recognized as being "entirely identical" with the classical probability concept. Lagrange's probability interval is today called a confidence interval. It is based on the distribution of the pivot (3.21), which involves the sufficient statistic h only and varies monotonically with the parameter. The random variable is h, but the pivotal argument allows us to speak of the parameter p as if it were a random variable.

Returning to the case $k > 2$, Lagrange succeeds in finding a lower bound, M say, for the probability that $|\delta_i| < \rho$ for all i, and he concludes that

$$P\left(h_i - \frac{\rho}{\sqrt{n}} < p_i < h_i + \frac{\rho}{\sqrt{n}}, \text{ for all } i\right) > M(\rho, k, h_1, \ldots, h_k), \qquad (3.22)$$

M being independent of n. It follows that $h_i - p_i$ tends in probability to zero for all i.

Lagrange stops here without reverting to his original problem about the mean. However, using the fact that

$$|(\mu - \overline{x}_n)\sqrt{n}| = |\sum x_i \delta_i| \leq \sum |x_i||\delta_i|$$

$$\leq \rho \sum |x_i| \text{ if } |\delta_i| \leq \rho \text{ for all } i,$$

it follows from (3.22) that $\overline{x}_n - \mu$ converges in probability to zero.

The method of statistical inference implied by Lagrange's procedure was overlooked by his contemporaries, perhaps because Laplace [148] independently had proposed to solve the inferential problem by the method of inverse probability.

Formula (3.16) gives an approximation to the likelihood function. However, setting $h_i = p_i + d_i/n$, the same method of proof gives an approximation to the sampling distribution, which is obtained by replacing the hs in (3.16) by the corresponding ps, as shown above for $k = 2$. This is the generalization of de Moivre's result.

When K. Pearson in the 1920s lectured on the history of statistics, he ([212], pp. 596–603) discovered Lagrange's result and remarked that it was the basis for his [203] χ^2 goodness-of-fit test.

3.3 De Morgan's Continuity Correction, 1838

Augustus de Morgan (1806–1871) improves de Moivre's approximation by introducing a "continuity correction" ([187], p. 77) based on the idea that each binomial probability should be interpreted as an area with unit base, which means that d in (3.1.13) and (3.1.14) should be replaced by $d + 1$.

J.V. Uspensky (1883–1946) writes ([262], p. 119):

> When we use an approximate formula instead of an exact one, there is always this question to consider: How large is the committed error? If, as is usually done, this question is left unanswered, the derivation of Laplace's formula [de Moivre's approximation supplemented by a term for the skewness] becomes an easy matter. However, to estimate the error comparatively long and detailed investigation is required.

He provides such an investigation, taking the continuity correction into account, and finds an upper limit for the absolute value of the error, which is of the order of n^{-1}, provided $npq \geq 25$; see ([262], p. 129).

4

Bayes's Posterior Distribution of the Binomial Parameter and His Rule for Inductive Inference, 1764

4.1 The Posterior Distribution of the Binomial Parameter, 1764

The English physician and philosopher David Hartley (1705–1757), founder of the Associationist school of psychologists, discusses some elementary applications of probability theory in his *Observations on Man* [118]. On the limit theorems he writes (pp. 338–339):

> Mr. *de Moivre* has shown, that where the Causes of the Happening of an Event bear a fixed Ratio to those of its Failure, the Happenings must bear nearly the same Ratio to the Failures, if the Number of Trials be sufficient; and that the last Ratio approaches to the first indefinitely, as the number of Trials increases. This may be considered as an elegant Method of accounting for that Order and Proportion, which we every-where see in the Phænomena of Nature. [...]
>
> An ingenious Friend has communicated to me a Solution of the inverse Problem, in which he has shewn what the Expectation is, when an event has happened p times, and failed q times, that the original Ratio of the Causes for the Happening or Failing of an Event should deviate in any given Degree from that of p to q. And it appears from this Solution, that where the Number of Trials is very great, the Deviation must be inconsiderable: Which shews that we may hope to determine the Proportions, and, by degrees, the whole Nature, of unknown Causes, by a sufficient Observation of their Effects.

This is a surprisingly clear statement of the law of large numbers for binomially distributed observations, based on direct and inverse probability, respectively.

We believe, as do most other commentators, that the ingenious friend was Bayes, who was the first to consider the probability of success, p say, as a uniformly distributed continuous random variable, so the statement above means that p converges in (posterior) probability to the observed relative frequency as the number of observations tends to infinity.

De Moivre, Bayes, and Hartley were all Fellows of the Royal Society so Hartley had first-hand access to both direct and inverse probability.

Hartley's formulation is remarkable also in two other respects. First, he uses the term "inverse problem" for the problem of finding probability limits for p. Second, he uses the terms from the ongoing philosophical discussions on the relation between cause and effect. De Moivre writes about design and chance, that is, the physical properties of the game and the probability distribution of the outcomes; he does not use the terms cause and effect. However, Hartley's terminology was accepted by many probabilists, who created a "probability of causes," also called inverse probability until about 1950 when Bayesian theory became the standard term.

To prevent misunderstandings of Hartley's unfortunate terminology de Morgan ([187], p. 53) explains:

> By a *cause*, is to be understood simply a state of things antecedent to the happening of an event, without the introduction of any notion of agency, physical or moral.

Thomas Bayes (c.1701–1761) was the son of a Presbyterian minister. He studied theology at Edinburgh University and afterwards became his father's assistant in London. In 1731 he became Presbyterian minister in Tunbridge Wells, southeast of London. He was unmarried, and after having inherited from his father in 1746, he became a wealthy man. He retired from the ministry in 1752, but kept living in Tunbridge Wells until his death. He seems to have led a quiet life, mainly occupied with his scholarly interests, beginning with theology, moving to mathematics and the natural sciences, and ending with statistical inference. He was elected a Fellow of the Royal Society in 1742.

When Bayes died in 1761 his relatives asked Richard Price (1723–1791), another Presbyterian minister, to examine the mathematical papers left by Bayes. Price found a paper on Stirling's formula and the paper "An Essay Towards Solving a Problem in the Doctrine of Chances," which he got published in two parts in the *Phil. Trans.* ([226], [227]) with introductory letters, comments, and extensions by himself.

Bayes's mathematics is correct, but his verbal comments are obscure and have caused much discussion, which recently has led to a new interpretation of his criterion for the application of his rule for inductive inference.

De Moivre had defined the expectation E of a game or a contract as the value V of the sum expected times the probability P of obtaining it, so $P = E/V$. Bayes chooses the value of an expectation as his primitive concept and defines probability as E/V. This is a generalization of the classical concept because an expectation may be evaluated objectively or subjectively. He then shows how the usual rules of probability calculus can be derived from this concept.

De Moivre had proved that

$$P(AB) = P(A)P(B|A) = P(B)P(A|B), \tag{4.1}$$

noting that the probability of the happening of both events equals the probability of the happening of one of them times the probability of the other, given that the first has happened. Bayes considers the two events as ordered in time and proves that for two "subsequent events," A occurring before B, we have

$$P(A|B) = P(AB)/P(B), \ P(B) > 0. \tag{4.2}$$

Bayes envisages a two-stage game of chance. At the first stage a real number p is chosen at random in the unit interval, and at the second stage n binomial trials are carried out with p as the probability of success. He describes how this game may be carried out by throwing balls at random on a rectangular table.

Denoting the probability of success in a single trial by $P(S) = p$, the probability of a successes in n independent trials is

$$P(S_n = a| \ p) = \binom{n}{a} p^a q^b, \ a + b = n, \ q = 1 - p, \ a = 0, 1, \dots, n.$$

By means of (4.1), Bayes gets the joint distribution of p and S_n,

$$P[(p_1 < p < p_2) \text{ and } (S_n = a)] = \int_{p_1}^{p_2} \binom{n}{a} p^a q^b dp, \ 0 \le p_1 < p_2 \le 1. \tag{4.3}$$

Integration from 0 to 1 gives the marginal distribution of S_n

$$P(S_n = a) = 1/(n+1), \ a = 0, 1, \dots, n. \tag{4.4}$$

Using (4.2) he gets the conditional distribution of p for given S_n

$$P(p_1 < p < p_2 | S_n = a) = \frac{(n+1)!}{a!b!} \int_{p_1}^{p_2} p^a q^b dp, \ p = P(S), \tag{4.5}$$

which is his final result, a distribution today called the beta distribution. Bayes remarks that the solution is uncontroversial under the conditions stated.

He has thus shown that probabilistic induction is possible for the physical experiment in question; all the probabilities involved have a frequency interpretation.

4.2 Bayes's Rule for Inductive Inference, 1764

In a scholium Bayes ([3], pp. 392–394) then asks whether "The same rule [our (4.5)] is the proper one to be used in the case of an event concerning the probability of which we absolutely know nothing antecedently to any trials made concerning it." He calls such an event an "unknown event" and formulates the problem as follows.

Given the number of times in which an unknown event has happened and failed: *Required* the chance that the probability of its happening in a single trial lies somewhere between any two degrees of probability that can be named.

Let us denote the unknown event by U and let U_n be the number of times U happens in n independent trials under the same circumstances. Hence, U corresponds to S and U_n to S_n, but S is not an unknown event because we know that $P(S)$ is uniformly distributed on $[0,1]$. That (4.1.5) is the proper rule to be used for finding limits for $P(U)$ seems, according to Bayes, to appear from the following consideration: "that concerning such an event I have no reason to think that, in a certain number of trials, it should rather happen any one possible number of times than another."

The statistical community has for nearly 200 years interpreted Bayes's postulate of ignorance as relating to the uniform distribution of $P(U)$. However, a closer reading of the quotation above has recently led to the result that Bayes refers to the uniform distribution of U_n. Note that $P(U)$ is unobservable and that we have only one observation of U_n so it is impossible to test the hypothesis about $P(U_n)$. For a survey of this discussion we refer to Stigler ([250] pp. 122–130).

Bayes's rule for inductive inference from n binomial trials may be summarized as follows. If we have no reason to think that U_n is not uniformly distributed on $(0, 1, \ldots, n)$, then limits for $P(U)$ may be calculated from the formula

$$P(p_1 < P(U) < p_2 | U_n = a) = \frac{(n+1)!}{a!b!} \int_{p_1}^{p_2} p^a q^{n-a} dp, \ 0 \le p_1 < p_2 \le 1,$$

(4.6)

which depends on the supposition that $P(U)$ is uniformly distributed on $[0, 1]$.

Thus ends the inferential part of Bayes's paper. He does not discuss where to find unknown events in nature; his paper contains no philosophy of science, no examples, and no data.

Price [226] attempts to remedy this defect in his commentary. As examples he discusses the drawings from a lottery and the probability of a sunrise tomorrow. Recognizing that Bayes's criterion of ignorance cannot be applied to himself regarding sunrises he invents "a person just brought forward into this world, knowing nothing at all about this phenomena." He (p. 410) concludes that "It should be carefully remembered that these deductions [about $P(U)$] suppose a previous total ignorance of nature." This implies that in the natural sciences "unknown events" are the exception rather than the rule. Usually we know something about the probability of a phenomenon under investigation and Bayes's rule is therefore seldom applicable. On this background it is no wonder that the Essay did not evoke any response from British mathematicians and natural scientists.

In the second part of the Essay [4] Bayes and Price derive an excellent approximation to the beta probability integral with limits of integration symmetric about the mode. Bayes's idea is to approximate the skew beta density with parameters (a, b) by the corresponding symmetric density with parameter $(a + b)/2$ and to introduce a measure of skewness on which the accuracy of the approximation depends. He obtains an asymptotic expansion, which is improved by Price. Finally, Price considers the expansion for $a + b \to \infty$ and obtains a series that is the expansion of the normal probability integral, but he does not recognize it as such. Also this part of the Essay was overlooked. For the details of the proofs we refer to Hald [113].

STATISTICAL INFERENCE BY INVERSE PROBABILITY. Inverse Probability from Laplace (1774), and Gauss (1809) to Edgeworth (1909)

5

Laplace's Theory of Inverse Probability, 1774–1786

5.1 Biography of Laplace

Pierre Simon Laplace (1749–1827) was born into a middle-class family at a small town in Normandy, where he spent his first 16 years. His father destined him for an ecclesiastical career and sent him to the University of Caen, where he matriculated in the Faculty of Arts with the intention to continue in the Faculty of Theology. However, after two years of study he left for Paris in 1768 bringing along a letter of recommendation from his mathematics teacher to d'Alembert. After having tested his abilities, d'Alembert secured him a post as teacher of mathematics at the École Militaire. He lived in Paris for the rest of his life.

Between 1770 and 1774 Laplace produced an exceptionally large number of important papers on mathematics, astronomy, and probability. In 1773, at the age of 24, he was elected an adjunct member of the Paris Academy of Sciences; he became a full member in 1785 and continued to be among the leading members of the Academy, of the succeeding Institut de France from 1795, and of the restored Academy from 1816. He carried out his scientific work during the old regime, the revolution, the Napoleonic era, and the restoration. He became a professor of mathematics at the École Normale and the École Polytechnique, a member of many government commissions, among them the commission on weights and measures, and a member of the Bureau des Longitudes.

Bonaparte made Laplace (an unsuccessful) Minister of the Interior for a period of six weeks and then a member of the Senate of which he later became Chancellor. After the restoration Louis XVIII created him a peer of France as a marquis. The various regimes used his reputation as an illustrious scientist to their own advantage, and Laplace used his influence to provide economic support for his research projects, for example, the Bureau des Longitudes, and for his scientific protégés. His adaptation to the various political systems has later been criticized.

Most of Laplace's contributions to mathematics were motivated by problems in the natural sciences and probability. To mention a few examples: celestial mechanics led him to study differential equations; problems in probability theory led him to difference equations, generating functions, Laplace transforms, characteristic functions, and asymptotic expansion of integrals.

In the early period of probability theory problems were usually solved by combinatorial methods. Lagrange and Laplace formulated the old problems as difference equations and developed methods for their solution; see Hald ([113] pp. 437–464). This is the reason why Laplace speaks of the analytical theory of probability in contradistinction to the combinatorial.

Besides his main interests in astronomy and probability, Laplace worked in physics and chemistry. He collaborated with Lavoisier about 1780 and with the chemist Berthollet from 1806. They were neighbors in Arcueil, where they created "The Society of Arcueil" as a meeting place for young scientists working in mathematics, physics, and chemistry; see Crosland [34].

In 1796 Laplace published the *Exposition du Systéme du Monde*, a popular introduction to his later monumental work *Traité de Méchanique Céleste* in four volumes 1799–1805 [154]. A fifth volume was published in 1825 [163].

After having completed his astronomical work in 1805, he resumed work on probability and statistics and published the *Théorie Analytique des Probabilités* (TAP), [159], the most influential book on probability and statistics ever written. In [160] he added the *Essai Philosophique sur les Probabilités* as a popular introduction to the second edition of the TAP. The *Essay* was also published separately and he kept on revising and enlarging it until the sixth edition. A third edition of the TAP, including important Supplements, appeared in 1820, and a fourth Supplement was added in 1825.

Among Laplace's numerous contributions to probability theory and statistics there are several outstanding ones: (1) A theory of statistical inference and prediction based on inverse probability [148]; (2) the asymptotic normality of posterior distributions [150], which may be called the central limit theorem for inverse probability; and (3) the asymptotic normality of the sampling distribution for sums of independent and identically distributed random variables ([155],[156], [159]), the central limit theorem for direct probability. He thus created a large-sample theory for both modes of probability.

Stigler [251] has translated Laplace's revolutionary paper "Memoir on the Probability of Causes of Events " [148] into English with an introduction that ends as follows.

> The influence of this memoir was immense. It was from here that "Bayesian" ideas first spread through the mathematical world, as Bayes's own article, [3], was ignored until after 1780 and played no important role in scientific debate until the twentieth century (Stigler [249]). It was also this article of Laplace's that introduced the mathematical techniques for the asymptotic analysis of posterior distributions that are still employed today. And it was here that the earliest

example of optimum estimation can be found, the derivation and characterization of an estimator that minimized a particular measure of posterior expected loss. After more than two centuries, we mathematical statisticians cannot only recognize our roots in this masterpiece of our science, we can still learn from it.

5.2 The Principle of Inverse Probability and the Symmetry of Direct and Inverse Probability, 1774

In the "Memoir on the Probability of Causes of Events" Laplace [148] begins by discussing direct and inverse probability by means of the urn model. He distinguishes between chance events, the outcome of drawings from an urn, and causes of events, the ratio of white to black tickets in the urn. If the cause is known and the event is unknown then the (direct) probability of the event can be found either by means of classical combinatorial methods or by Laplace's analytical methods. If the event is known and the cause is unknown then a new principle for finding the probability of the cause is needed. Laplace formulates the principle of inverse probability as follows.

> If an event can be produced by a number n of different causes, the probabilities of the existence of these causes given the event are to each other as the probabilities of the event given the causes, and the probability of the existence of each of them is equal to the probability of the event given that cause divided by the sum of all the probabilities of the event given each of the causes.

Laplace does not offer any reason for this principle; obviously he considers it intuitively reasonable.

Denoting the n mutually exclusive and exhaustive causes by $C_1, \ldots, C_n$ and the event by E and using the modern notation for conditional probability Laplace thus considers the following scheme.

$$
\begin{array}{ll}
\text{Causes } (n \text{ urns}) & C_1, \ldots, C_n \\
\text{Direct probability} & P(E|C_1), \ldots, P(E|C_n) \\
\text{Inverse probability} & P(C_1|E), \ldots, P(C_n|E)
\end{array}
$$

Direct probability corresponds to probabilistic deduction and inverse probability to probabilistic induction.

It is a remarkable fact that Laplace considers conditional probabilities only. His principle amounts to the symmetry relation

$$P(C_i|E) \propto P(E|C_i), \ i = 1, \ldots, n,$$

which is the form he ordinarily uses. His intuitive reasoning may have been as follows: If the probability of the observed event for a given cause is large relative to the other probabilities then it is relatively more likely that the event has been produced by this cause than by any other cause.

Applying the principle to parametric statistical models he uses the symmetry relation for the frequency functions in the form

$$p(\underline{\theta}|\underline{x}) \propto f(\underline{x}|\underline{\theta}), \ \underline{x} = (x_1, \ldots, x_n), \ \underline{\theta} = (\theta_1, \ldots, \theta_m); \tag{5.1}$$

that is, the posterior density of $\underline{\theta}$ for given $\underline{x}$ is proportional to the density of $\underline{x}$ for given $\underline{\theta}$.

In 1774 Bayes's paper, [3], was not known among French probabilists. However, by 1781 Laplace knew Bayes's paper and this may have induced him to derive his principle from a two-stage model with equal probabilities for the causes. In 1786 he points out that the theory of inverse probability is based on the relation

$$P(C_i|E) = P(C_iE)/P(E),$$

and assuming that $P(C_i) = 1/n, \ i = 1, \ldots, n,$ he finds

$$P(C_i|E) = \frac{P(E|C_i)}{\sum P(E|C_i),} \tag{5.2}$$

in agreement with his 1774 principle.

It is thus clear that at least from 1786 on Laplace's principle had two interpretations: a frequency interpretation based on a two-stage model with objective probabilities, and an interpretation based on the principle of insufficient reason, also called the principle of indifference. This distinction is clearly explained by Cournot ([33], Chapter 8), who notes that the first interpretation is unambiguous and uncontestable, whereas the second is subjective and arbitrary.

The proof above is reproduced in the first edition of the TAP ([159], II, § 1). In the second edition ([160], II, § 1) Laplace introduces a nonuniform distribution of causes, and replacing $1/n$ by $P(C_i)$ in (5.2) he obtains

$$P(C_i|E) = \frac{P(C_i)P(E|C_i)}{\sum P(C_i)P(E|C_i)}, \ i = 1, \ldots, n, \tag{5.3}$$

which today is called Bayes's formula.

Laplace had previously discussed cases of nonuniform priors for parametric models in the form

$$p(\theta|\underline{x}) \propto f(\underline{x}|\theta)w(\theta), \tag{5.4}$$

where $w(\theta)$ denotes the prior distribution. He remarks that if causes are not equally probable then they should be made so by subdividing the more probable ones, just as events having unequal probabilities may be perceived as unions of events with equal probabilities. This is easily understood for a finite number of urns with rational prior probabilities. In the continuous case Laplace ([151], § 35) uses this idea only in the trivial form

$$P(\theta_1 < \theta < \theta_2 | \underline{x}) = \int_{\theta_1}^{\theta_2} g(\underline{x}|\theta)d\theta, \ g(x|\theta) = p(\underline{x}|\theta)w(\theta),$$

expressing the asymptotic expansion of the integral in terms of the maximum value of $g(\underline{x}|\theta)$ and the derivatives of $\ln g(\underline{x}|\theta)$ at this value.

In the theory of inverse probability it became customary tacitly to assume that causes, hypotheses, or parameters are uniformly distributed, unless it is known that this is not so.

In the specification of the statistical model (5.1) it is tacitly understood that $\underline{x}$ and $\underline{\theta}$ belong to well-defined spaces, the sample space and the parameter space, respectively.

Inasmuch as $p(\underline{\theta}|\underline{x})$ is a probability distribution it follows from (5.3) and (5.4) that

$$p(\underline{\theta}|\underline{x}) = f(\underline{x}|\underline{\theta})w(\underline{\theta})/\int f(\underline{x}|\underline{\theta})w(\underline{\theta})d\underline{\theta}. \tag{5.5}$$

The revolutionary step taken by Laplace in 1774 is to consider scientific hypotheses and unknown parameters as random variables and a par with observations. As noted by Cournot and many others there is no empirical evidence for this supposition; nature does not select parameters at random.

A note on inverse probability and mathematical likelihood. In (5.1) the random variable $\underline{x}$ is observable and $f(\underline{x}|\theta)$ has a frequency interpretation for a given value of the parameter θ, which is an unknown constant. Hence, all information on θ is contained in the observed value of $\underline{x}$ and the statistical model $f(\underline{x}|\theta)$ that links the two together. The inference problem is to find limits for θ. At the times of Laplace the only numerical measure of uncertainty was probability, so even though Laplace considered θ as an unknown constant, he had in some way to introduce a probability distribution for θ. He chose the simplest possible solution to this mathematical problem by introducing a fictitious random variable, uniformly distributed on the parameter space, and linking it to $f(\underline{x}|\theta)$ by means of the relation (5.1). This is an ingenious mathematical device to reach his goal. It is clear that the mathematical properties of $f(\underline{x}|\theta)$ as a function of θ carry over to $p(\theta|\underline{x})$; for example the posterior mode $\hat{\theta}$ equals the value of θ maximizing $f(\underline{x}|\theta)$, today called the maximum likelihood estimate.

To clear up the confusion connected with the interpretation of (5.1) Fisher [66] proposed to call any function of θ, proportional to $f(\underline{x}|\theta)$ the likelihood function, the constant of proportionality being arbitrary. Hence,

$$L(\theta) = L(\theta|\underline{x}) \propto f(\underline{x}|\theta) \tag{5.6}$$

is not a probability distribution; there is no normalizing constant involved as in (5.5).

5.3 Posterior Consistency and Asymptotic Normality in the Binomial Case, 1774

Bernoulli had proved that $h = x/n$, where x is binomial (n, θ), converges in (direct) probability to θ, so to justify the principle of inverse probability, Laplace [148] naturally wanted to prove that θ converges in (inverse) probability to h.

By means of (5.1), Laplace gets the posterior distribution

$$p(\theta|n, h) = \frac{1}{B(x+1, n-x+1)} \theta^x (1-\theta)^{n-x}, \qquad (5.7)$$

for $x = 0, 1, \ldots, n$, $0 \le \theta \le 1$, which is the beta distribution with parameters $(x+1, n-x+1)$.

He then proposes to show that

$$P_n = P(|\theta - h| < \varepsilon | n, h) \to 1, \text{ as } n \to \infty,$$

where ε " can be supposed less than any given quantity." Setting

$$\varepsilon = n^{-\delta}, \ 1/3 < \delta < 1/2,$$

he proves that

$$P_n = \int_{h-\varepsilon}^{h+\varepsilon} p(\theta|n, h)d\theta \sim \Phi[\varepsilon\sqrt{n/h(1-h)}] - \Phi[-\varepsilon\sqrt{n/h(1-h)}], \quad (5.8)$$

which tends to 1 because $\varepsilon\sqrt{n} \to \infty$.

The proof is the first instance of Laplace's method of asymptotic expansion of definite integrals. By means of Taylor's formula he expands the logarithm of the integrand into a power series around its mode, which in the present case is h. For $\theta = h + t$ he finds

$$\ln p(\theta|n, h) = \ln p(h|n, h) - \frac{1}{2}nt^2(h^{-1} + k^{-1}) + \frac{1}{3}nt^3(h^{-2} + k^{-2}) - \cdots, \ k = 1-h.$$
$$(5.9)$$

He evaluates the constant term by means of Stirling's formula which gives

$$p(h|n, h) \sim \sqrt{n/2\pi hk}. \qquad (5.10)$$

He remarks that $|t| < \varepsilon$ in the evaluation of (5.8) which means that terms of order 3 or more are negligible in (5.9). Hence,

$$P_n \sim \frac{\sqrt{n}}{\sqrt{2\pi hk}} \int_{-\varepsilon}^{\varepsilon} \exp(-nt^2/2hk)dt, \qquad (5.11)$$

which leads to (5.8).

He completes this result by giving the first proof of the fact that the integral of the normal density function equals 1. His proof is somewhat artificial and he later ([149], Art. 23) gave a simpler proof by evaluating the double integral

$$\int_0^\infty \int_0^\infty \exp[-s(1+x^2)]dsdx = \int_0^\infty (1+x^2)^{-1}dx = \frac{1}{2}\pi,$$

and using the transformations $s = u^2$ and $sx^2 = t^2$ to show that the integral equals

$$2\int_0^\infty \exp(-u^2)du \int_0^\infty \exp(-t^2)dt.$$

Finally, he evaluates the tail probability of $p(\theta|n,h)$ to get a formula by which "we can judge the error made by taking $E = 1$." $[P_n = E]$. Introducing

$$y(t) = \frac{\ln p(h+t|n,h)}{\ln p(h|n,h)} = \left(1+\frac{t}{h}\right)^{nh}\left(1-\frac{t}{k}\right)^{nk},$$

he gets for the right tail that

$$\int_\varepsilon^k y(t)dt = y(\varepsilon)\int_0^{k-\varepsilon} \exp[\ln y(t+\varepsilon) - \ln y(\varepsilon)]dt$$

$$\sim y(\varepsilon)\int_0^{k-\varepsilon} \exp(-n\varepsilon t/hk)dt,$$

which equals $y(\varepsilon)hk/(n\varepsilon)$. This is the first instance of his method for evaluating tail probabilities. It follows that

$$P_n \sim 1 - \frac{\sqrt{hk}}{\varepsilon\sqrt{2\pi n}}[y(-\varepsilon) + y(\varepsilon)], \tag{5.12}$$

which is easy to calculate.

It is clear that Laplace's proof implies that θ is asymptotically normal $(h, hk/n)$, [see (5.11)], although Laplace does not discuss this result in the present context. This section of the 1774 paper is a remarkable achievement. The 24-year-old Laplace had on a few pages given a simple proof of the inverse of Bernoulli's and de Moivre's complicated proofs, which had taken these authors 20 and 12 years, respectively, to produce.

In his 1781 and 1786 papers he improves (5.8) and (5.12) by taking more terms of the expansions into account.

He uses the asymptotic normality of θ to calculate credibility intervals for θ in the form $h \pm u\sqrt{hk/n}$, where u is normal $(0,1)$.

He tests the hypothesis $\theta \leq r$, say, against $\theta > r$ by comparing the tail probability $P(\theta \leq r)$ with its complement. For two independent binomial samples, he ([151], Art. 40) proves that $\theta_1 - \theta_2$ is asymptotically normal with mean $h_1 - h_2$ and variance $h_1k_1/n_1 + h_2k_2/n_2$, which he uses for testing the hypothesis $\theta_2 \leq \theta_1$ against $\theta_2 > \theta_1$. He thus laid the foundation for the theory of testing statistical hypotheses.

5.4 The Predictive Distribution, 1774–1786

Let E_1 and E_2 be two conditionally independent events so that

$$P(E_1 E_2 | C_i) = P(E_1 | C_i) P(E_2 | C_i), \quad i = 1, 2, \ldots, n,$$

and let $P(C_i) = 1/n$. The probability of the future event E_2, given that E_1 has occurred, equals

$$P(E_2 | E_1) = \sum P(E_1 | C_i) P(E_2 | C_i) / \sum P(E_1 | C_i) \tag{5.13}$$
$$= \sum P(E_2 | C_i) P(C_i | E_1).$$

Comparing with

$$P(E_2) = \sum P(E_2 | C_i) P(C_i),$$

it will be seen that the conditional probability of E_2, given E_1, is obtained from the unconditional probability by replacing the prior distribution of C_i by the updated prior, given E_1. For the continuous case we have similarly

$$P(E_2 | E_1) = \int P(E_2 | \theta) P(\theta | E_1) d\theta. \tag{5.14}$$

This is the basic principle of Laplace's theory of prediction, developed between 1774 and 1786. He uses (5.13) for finding the probability of a future series of events and for updating the prior successively.

Let E_1 be the outcome of $n = a + b$ binomial trials with a successes and E_2 the outcome of $m = c + d$ trials under the same circumstances. Hence,

$$p(c|a) = \binom{m}{c} \int_0^1 \theta^{a+c} (1 - \theta)^{b+d} d\theta \Big/ \int_0^1 \theta^a (1 - \theta)^b d\theta \tag{5.15}$$

$$= \binom{m}{c} \frac{(a+1)^{[c]} (b+1)^{[d]}}{(n+2)^{[m]}} \tag{5.16}$$

$$\cong \binom{m}{c} \frac{(a+c)^{a+c+1/2} (b+d)^{b+d+1/2} n^{n+3/2}}{a^{a+1/2} b^{b+1/2} (n+m)^{n+m+3/2}}, \quad c = 0, 1, \ldots, m; \tag{5.17}$$

see Laplace ([149], Art. 17). We have used the notation $a^{[x]} = a(a+1) \cdots (a+x-1)$. In the form (5.16) the analogy to the binomial is obvious. Laplace obtains (5.17) from (5.16) by means of Stirling's formula. Formula (5.15) is the beta-binomial or the inverse hypergeometric distribution which is also known in the form

$$p(c|a) = \binom{a+c}{c} \binom{b+d}{d} \Big/ \binom{n+m+1}{m}.$$

Laplace derives similar results without the binomial coefficient in the 1774 paper where he considers a specified sequence of successes and failure.

To find a large-sample approximation to $p(c|a)$ based on (5.17), Laplace ([159], II, Art. 30) keeps $h = a/n$ fixed, as $n \to \infty$. Assuming that m is large, but at most of the order of n, he proves that c is asymptotically normal with mean mh and variance

$$mh(1 - h)(1 + \frac{m}{n}).$$

Hence, Laplace warns against using the binomial with mean mh and variance $mh(1 - h)$ for prediction unless m/n is small.

Laplace ([151], Art. 41–43) generalizes the asymptotic theory further by evaluating the integral

$$E(z^r(\theta)|a) = \int_0^1 z^r(\theta)\theta^a(1 - \theta)^b d\theta \Big/ \int_0^1 \theta^a(1 - \theta)^b d\theta,$$

which gives the conditional probability for the r-fold repetition of a compound event having probability $z(\theta)$.

Of particular interest is the so-called rule of succession which gives the probability of a success in the next trial, given that a successes have occurred in the first n trials. Setting $m = c = 1$ this probability becomes $(a+1)/(n+2)$. It is clear that $p(c|a)$ can be found as the product of such successive conditional probabilities times a binomial coefficient.

Laplace ([149], Art. 33) also derives the rule of succession for multinomial variables.

Setting $E_1 = (x_1, \ldots, x_n)$ and $E_2 = x_{n+1}$ in (5.14) we get the predictive density

$$p(x_{n+1}|x_1, \ldots, x_n) = \int f(x_{n+1}|\theta)p(\theta|x_1, \ldots, x_n)d\theta, \tag{5.18}$$

where $f(x|\theta)$ is the density of x and $p(\theta|x_1, \ldots, x_n)$ is given by (5.11).

5.5 A Statistical Model and a Method of Estimation. The Double Exponential Distribution, 1774

At the time of Laplace [148] it was customary in the natural sciences to use the arithmetic mean of repeated observations under essentially the same circumstances as the estimate of the true value of the phenomenon in question, but a general probabilistic foundation for this practice was lacking. It was the aim of Laplace to prove that $|\bar{x} - \theta| \to 0$ in probability as $n \to \infty$, just as he had justified the use of the relative frequency as an estimate of the binomial parameter.

Errors were expressed as $|x - \theta|$ and error distributions were considered as symmetric about zero with finite range; only the rectangular, the triangular, and the semicircular distributions had been proposed before Laplace. Thomas Simpson (1710–1761), [237], had derived the sampling distribution of the mean for observations from a symmetric triangular distribution, a rather complicated function, and had shown numerically that $P(|\bar{x}| < k) > P(|x_1| < k)$ for

$n = 6$ and two values of k, from which he concluded that it is advantageous to use the mean as the estimate of θ.

As a first step Laplace [148] introduces a new error distribution with infinite support, the double exponential distribution

$$f(x|\theta, m) = \frac{m}{2}e^{-m|x-\theta|}, \quad -\infty < x < \infty, \ -\infty < \theta < \infty, \ 0 < m < \infty.$$

It follows from the principle of inverse probability that

$$p(\theta, m|\underline{x}) \propto (m/2)^n \exp(-m \sum |x_i - \theta|).$$

Next Laplace proposes two principles for estimating the location parameter. According to the first, the estimate $\tilde{\theta}$ should be chosen such that it is equally probable for the true value to fall below or above it; that is, $\tilde{\theta}$ is the posterior median. According to the second principle the estimate should minimize the expected error of estimation. He proves that the two principles lead to the same estimate.

Here we have for the first time a completely specified statistical model and a well-defined principle of estimation.

It is remarkable that Laplace starts from the joint distribution of θ and m although he is interested only in estimating θ. He could have limited himself to the case with a known scale parameter as customary by other authors at the time. Now he had to take the nuisance parameter m into account.

Another remarkable fact is the apparent mathematical simplicity that he obtains by keeping strictly to the then prevailing error concept. The observational error is expressed as $|x - \theta|$ and the error of estimation as $|\theta - \tilde{\theta}|$.

Laplace does not comment on the fact that he uses the posterior mode as the estimate of the binomial parameter but the posterior median as the estimate of the location parameter. Because the median and the mode coincide for a symmetric distribution he needed an argument for choosing between them. This he found by likening estimation to a game of chance in which the player's expected loss should be minimized. However, he does not explain why a scientists's "loss" caused by an error of estimation should be measured in the same way as a player's pecuniary loss.

It turned out that the solution of the estimation problem for an arbitrary sample size is extremely difficult so Laplace limits himself to consider the case $n = 3$.

Let the observations be labeled such that $x_1 \leq x_2 \leq x_3$. Laplace makes the transformation

$$\lambda = \theta - x_1, \ a_1 = x_2 - x_1, \ a_2 = x_3 - x_2,$$

so the posterior distribution becomes

$$p(\lambda, m|\underline{x}) \propto (m^3/8)\exp[-m(|\lambda| + |a_1 - \lambda| + |a_1 + a_2 - \lambda|).\tag{5.19}$$

Obviously, the marginal distribution of m, $h(m|a_1, a_2)$ say, depends only on (a_1, a_2), which are ancillary with respect to λ. The conditional distribution of λ, given m and $\underline{x}$ is denoted by $g(\lambda|a_1, a_2, m)$, so

$$p(\lambda, m|\underline{x}) = g(\lambda|a_1, a_2, m)h(m|a_1, a_2).$$

Hence, for a given value of m we have

$$g(\lambda|a_1, a_2, m) \propto p(\lambda, m|\underline{x}).$$

Assuming first that m is known, Laplace estimates λ by setting the integral of (5.19) from $-\infty$ to $\tilde{\lambda}$ equal to half the integral from $-\infty$ to ∞. Solving for $\tilde{\lambda}$ he finds

$$\tilde{\lambda} = \tilde{\theta} - x_1 = a_1 + \frac{1}{m}\ln\left(1 + \frac{1}{3}e^{-ma_1} - \frac{1}{3}e^{-ma_2}\right), \quad a_1 > a_2.$$

This is the first disappointment: $\tilde{\theta}$ differs from the arithmetic mean. Laplace notes that

$$\lim_{m\to 0}\tilde{\theta} = x_1 + \frac{1}{3}(2a_1 + a_2) = \bar{x},$$

so the arithmetic mean is obtained only in the unrealistic case where the observed errors are uniformly distributed on the whole real line.

Remarking that m usually is unknown, Laplace proceeds to find $\tilde{\lambda}$ from the marginal density of λ using that

$$p(\lambda|a_1, a_2) = \int_0^\infty g(\lambda|a_1, a_2, m)h(m|a_1, a_2)dm, \tag{5.20}$$

where he obtains $h(m|a_1, a_2)$ by integration of (5.19), which gives

$$h(m|a_1, a_2) \propto m^2 e^{-m(a_1+a_2)}\left(1 - \frac{1}{3}e^{-ma_1} - \frac{1}{3}e^{-ma_2}\right).$$

He does not discuss how to use this result for estimating m.

Using (5.20) for solving the equation

$$\int_{-\infty}^{\tilde{\lambda}} p(\lambda|a_1, a_2)\,d\lambda - \frac{1}{2}\int_{-\infty}^\infty p(\lambda|a_1, a_2)d\lambda = 0,$$

he finds that $\tilde{\lambda}$ is the root of a polynomial equation of the fifteenth degree and proves that there is only one root smaller than a_1. This is the second disappointment: $\tilde{\lambda}$ differs from $\bar{x}$, and for $n > 3$ the solution is so complicated that it is of no practical value.

Stigler ([250], pp. 105–117) points out an error in Laplace's manipulations with the conditional probabilities involved and shows that the correct solution is found as the root of an equation of the third degree. Although Laplace's paper did not lead to practical results it is of great value because of its many new ideas.

After this fruitless attempt to solve the estimation problem in a fully specified model Laplace turned to the problem of estimating the coefficients in the linear model $y = X\beta + \varepsilon$, assuming only that the errors are symmetrically distributed around zero; see Section 7.4.

5.6 The Asymptotic Normality of Posterior Distributions, 1785

In his 1774 proof of the asymptotic normality of the beta distribution Laplace uses a uniform prior as he did later on in many other problems. However, these results may be considered as special cases of his [150] general theory based on an arbitrary differentiable prior.

First, he ([150], Art. 1; [159], I, Art. 23) generalizes the binomial to a multinomial model with cell probabilities depending on a parameter θ with prior density $w(\theta)$ so that

$$p(\theta|\underline{x}) \propto g_1^{x_1}(\theta) \cdots g_k^{x_k}(\theta)w(\theta), \tag{5.21}$$

where $\underline{x} = (x_1, \ldots, x_k)$, $\sum g_i(\theta) = 1$, and $\sum x_i = n$. He derives an asymptotic expansion of the corresponding probability integral for $h_i = x_i/n, i = 1, \ldots, k$, fixed and $n \to \infty$

Next, he ([150], Art. 6; [159], I, Art. 27) discusses the continuous case

$$p(\theta|\underline{x}) = f(\underline{x}|\theta)w(\theta)/p(\underline{x}), \quad \underline{x} = (x_1, \ldots, x_n), \tag{5.22}$$

where $f(x|\theta)$ is a frequency function, $f(\underline{x}|\theta) = f(x_1|\theta) \cdots f(x_n|\theta)$ and

$$p(\underline{x}) = \int f(\underline{x}|\theta)w(\theta)d\theta. \tag{5.23}$$

He derives an asymptotic expansion of the probability integral for $\underline{x}$ fixed and $n \to \infty$.

Laplace derives the first three terms of the asymptotic expansions; see Hald ([114], § 13.4). We here only indicate the method of proof and derive the main term. It is assumed that the posterior distribution is unimodal and disappears at the endpoints of its support. The mode is denoted by $\hat{\theta}$ and, like Laplace, we abbreviate $p(\theta|\underline{x})$ to $p(\theta)$. As in the binomial case Laplace evaluates the probability integral by developing the integrand into a power series, which he then integrates. For $\alpha < \hat{\theta} < \beta$ we have

$$\int_\alpha^\beta p(\theta)d\theta = p(\hat{\theta}) \int_\alpha^\beta \exp[\ln p(\theta) - \ln p(\hat{\theta})]d\theta \tag{5.24}$$

$$= p(\hat{\theta}) \int_a^b \exp(-u^2/2)\frac{d\theta}{du}du,$$

where
$$-u^2/2 = \ln p(\theta) - \ln p(\hat{\theta}), \tag{5.25}$$

and the limits of integration are
$$a = -[2\ln\{p(\hat{\theta})/p(\hat{\theta} - \alpha)\}]^{1/2}$$

and
$$b = [2\ln\{p(\hat{\theta})/p(\hat{\theta} + \beta)\}]^{1/2}.$$

By means of the transformation (5.25), the problem has thus been reduced to finding $d\theta/du$ as a function of u, which may be done by means of a Taylor expansion. We therefore introduce the coefficients
$$c_k = \frac{d^k \ln p(\theta)}{d\theta^k}\Big|_{\theta=\hat{\theta}}, \quad k = 1, 2, \ldots,$$

noting that $c_1 = 0, c_2 < 0$, and that $c_2, c_3, \ldots$ are $O(n)$. It follows that
$$u^2 = -c_2(\theta - \hat{\theta})^2 - \frac{1}{3}c_3(\theta - \hat{\theta})^3 + \ldots, \tag{5.26}$$

which by differentiation gives
$$\frac{d\theta}{du} = \pm \frac{1}{\sqrt{-c_2}}(1 + cu/\sqrt{-c_2} + \ldots),$$

where c depends on the ratio c_3/c_2 which is $O(1)$.

Using that the complete integral of $p(\theta)$ equals unity, $p(\hat{\theta})$ may be eliminated from (5.24) and we get
$$\int_\alpha^\beta p(\theta)d\theta \sim \int_a^b \phi(u)du,$$

which shows that u is asymptotically normal $(0, 1)$. From (5.26) it follows that
$$u = \pm\sqrt{-c_2}(\theta - \hat{\theta})[1 + (c_3/6c_2)(\theta - \hat{\theta}) + \ldots].$$

Hence, in a neighborhood of $\hat{\theta}$ of order $n^{-1/2}$, u is asymptotically linear in θ so that θ becomes asymptotically normal with mean $\hat{\theta}$ and variance given by
$$\frac{1}{-c_2} = \frac{1}{V(\theta)} = -\frac{d^2 \ln p(\hat{\theta})}{d\theta^2}. \tag{5.27}$$

By a more detailed analysis Laplace proves that
$$p(\theta)d\theta = \phi(u)[1 + a_1 u + a_2(u^2 - 1) + (a_3 u^3 - a_1 a_2 u) + \cdots]du, \tag{5.28}$$

where the as are expressed in terms of the cs and a_i is of order $n^{-i/2}$.

This is Laplace's fundamental ("central") limit theorem, which is the foundation for the large sample theory based on inverse probability. Modern versions have been discussed by Hald ([114], § 13.5). For the two-parameter model, Laplace ([159], I, Art. 28) shows that $(\hat{\theta}_1, \hat{\theta}_2)$ for large n is bivariate normal with mean (θ_1, θ_2) and inverse dispersion matrix equal to

$$(c_{ij}) = \left(-\frac{\partial^2 \ln p(\hat{\theta})}{\partial \hat{\theta}_i \partial \hat{\theta}_j}\right), \quad (i, j) = 1, 2, \dots. \tag{5.29}$$

We now note some consequences of Laplace's theorem. Because the first term of $\ln p(\theta)$ is of order n and the second term, $\ln w(\theta)$, is of order 1, it follows that $\hat{\theta}$ and $V(\theta)$ for large n are independent of the prior distribution. This property was presumably known by Laplace and his contemporaries; it is explicitly expressed by Cournot ([33], § 95), who remarks that in this case the posterior distribution assumes "an objective value, independent of the form of the unknown [prior] function."

In modern terminology the likelihood function $L(\theta)$ is defined as proportional to $f(\underline{x}|\theta)$ and $l(\theta) = \ln L(\theta)$. If the prior is uniform then

$$\frac{d}{d\theta} \ln p(\theta) = \frac{d}{d\theta} \ln f(\underline{x}|\theta) = \frac{d}{d\theta} l(\theta), \tag{5.30}$$

so that $\hat{\theta}$ equals the maximum likelihood estimate. Because of the equivalence of direct and inverse probability (see, e.g., the binomial case treated above) it follows intuitively that the maximum likelihood estimate is asymptotically normal with mean θ and that

$$1/V(\hat{\theta}) = 1/E[V(\hat{\theta}|\underline{x})] = E\left(-\frac{d^2 \ln f(\underline{x}|\theta)}{d\theta^2}\right),$$

a result formally proved by Edgeworth [47] under some restrictions.

Laplace's two proofs above are examples of his method of asymptotic expansions of definite integrals. In the TAP ([159], II, Art. 23) he gives a simpler proof in connection with a discussion of estimating the location parameter of a symmetric error distribution, looking only for the main term of the expansion and assuming that the prior is uniform. It is, however, easy to see that his proof holds also for the general model (5.22).

Expanding $\ln p(\theta)$ in Taylor's series around $\hat{\theta}$ he finds

$$\ln p(\theta) = \ln p(\hat{\theta}) + \frac{1}{2}\left(\theta - \hat{\theta}\right)^2 \frac{d^2}{d\hat{\theta}^2} \ln p(\hat{\theta}) + \cdots. \tag{5.31}$$

For values of $\theta - \hat{\theta}$ at most of order $n^{-1/2}$ the terms after the second can be neglected. Using that the complete integral of $p(\theta)$ equals unity, $p(\hat{\theta})$ is eliminated with the result that θ is asymptotically normal $[\hat{\theta}, V(\hat{\theta})]$. Laplace's proof, supplemented by regularity conditions, is the basis for the modern proofs.

6

A Nonprobabilistic Interlude: The Fitting of Equations to Data, 1750–1805

6.1 The Measurement Error Model

We consider the model

$$y_i = f(x_{i1}, \ldots, x_{im}; \beta_1, \ldots, \beta_m) + \varepsilon_i, \ i = 1, \ldots, n, \ m \leq n,$$

where the ys represent the observations of a phenomenon, whose variation depends on the observed values of the xs, the βs are unknown parameters, and the εs random errors, distributed symmetrically about zero. Denoting the true value of y by η, the model may be described as a mathematical law giving the dependent variable η as a function of the independent variables $x_1, \ldots, x_m$ with unknown errors of observation equal to $\varepsilon = y - \eta$.

Setting $\varepsilon_1 = \cdots = \varepsilon_n = 0$, we obtain for $n > m$ a set of inconsistent equations, called the equations of condition. Replacing the βs by the estimates $b_1, \ldots, b_m$, say, we get the adjusted values of the observations and the corresponding residuals $e_i = y_i - \tilde{y}_i, \ i = 1, \cdots, n$, small residuals indicating good estimates.

We call the model linear if f is linear in the βs. If f is nonlinear, it is linearized by introducing approximate values of the βs and using Taylor's formula. In the following discussion of the estimation problem it is assumed that linearization has taken place so that the reduced model becomes

$$y_i = \beta_1 x_{i1} + \cdots + \beta_m x_{im} + \varepsilon_i, \ i = 1, \ldots, n, \ m \leq n. \tag{6.1}$$

For one independent variable we often use the form

$$y_i = \alpha + \beta x_i + \varepsilon_i.$$

Using matrix notation (6.1) is written as $y = X\beta + \varepsilon$.

In the period considered no attempts were made to study the sampling distribution of the estimates. The efforts were concentrated on devising simple methods for finding point estimates of the parameters and on finding objective methods.

The first methods of estimation were subjective, as for example the method of selected points that consists of choosing m out of the n observations and solving the corresponding m equations of condition. Hence, m of the residuals are by definition equal to zero. Many scientists using this method calculated the remaining $n - m$ residuals and studied their sign and size to make sure that there were no systematic deviations between the observations and the proposed law. This procedure gave at the same time a qualitative check of the model, an impression of the goodness of fit which could be compared with previous experience on the size of observational errors, and a possibility for identifying outliers.

In problems of estimation it is assumed that the model is given, usually formulated by the scientist by means of a combination of theoretical and empirical knowledge.

In the following sections we sketch three methods of fitting linear equations to data: the method of averages, the method of least absolute deviations, and the method of least squares. A more detailed discussion, including the scientific background in astronomy and geodesy, the original data, and their analysis, is due to Stigler [250] and Farebrother [58], who carry the history up to about 1900.

In the period also a fourth method was developed: the method of minimizing the largest absolute residual, today called the minimax method. However, we do not explain this method because it is of no importance for our further discussion of methods of estimation. Farebrother [58] has given a detailed history of the method.

6.2 The Method of Averages by Mayer, 1750, and Laplace, 1788

The German cartographer and astronomer Tobias Mayer (1723–1763) studied the libration of the moon by making 27 carefully planned observations of the crater Manilius during a period of a year. Using spherical trigonometry he [174] derived a nonlinear relation between three measured arcs and three unobservable parameters. Linearizing this relation and inserting the observed values he obtained 27 equations of the form

$$\beta_1 - y_i = \beta_2 x_{i2} - \beta_3 x_{i3}, \quad i = 1, \ldots, 27,$$

where y_i is the observed latitude of Manilius in relation to the moon's apparent equator and $x_{i2}^2 + x_{i3}^2 = 1$, because x_{i2} and x_{i3} are the sine and the cosine of the same observed angle. The observations were planned to obtain a large variation of x_2 and thus also of x_1.

To estimate the parameters Mayer first uses the method of selected points. He chooses three of the 27 equations in such a way that large differences between the three values of x_2 are obtained with the purpose to get a good determination of the unknowns. He solves the three equations by successive elimination of the unknowns. However, he remarks that the method is unsatisfactory because selecting three other equations will lead to other estimates, hence all the observations should be used.

He proposes to divide the 27 equations into three groups of nine each, to sum the equations within each group, and to solve the resulting three equations. This method became known as Mayer's method; one may of course use averages instead of sums wherefore it later was called the method of averages.

Mayer states that "the differences between the three sums are made as large as possible," that is, the same principle as used in selecting the three equations above. To obtain this goal, he classifies the equations according to the size of x_{i2}, the nine largest values defining the first group and the nine smallest the second.

The method of averages became popular because of its simplicity, both conceptually and numerically. However, for more than one independent variable there may be several ways of obtaining large contrasts between the coefficients in the equations. Mayer did not face this problem because in his example the two independent variables are functionally related.

Laplace [152] generalizes Mayer's method in a paper in which he explains the long inequalities in the motions of Jupiter and Saturn as periodic with a period of about 900 years, based on 24 observations on the longitude of Saturn over the period 1591–1785. After linearization Laplace obtains 24 equations of the form

$$y_i = \beta_0 + \beta_1 x_{i1} + \beta_2 x_{i2} + \beta_3 x_{i3} + \varepsilon_i, \ i = 1, \ldots, 24,$$

where $x_{i2}^2 + x_{i3}^2 = 1$. To find the four unknowns from the 24 inconsistent equations obtained by setting the εs equal to zero, Laplace constructs four linear combinations of the equations. For simplicity he uses coefficients equal to $+1, 0$, and -1 only, so that the calculations are reduced to additions and subtractions. This is an important advance over Mayer's method, which is based on disjoint subsets of the equations.

The first linear combination is obtained by adding all 24 equations, the second by subtracting the sum of the last 12 from the sum of the first 12. Laplace does not explain the principle used for constructing the third and the fourth combination; he only lists the number of the equations to be added, subtracted, or neglected. However, it is easy to see from this list that the third combination is obtained, with one exception, from the 12 equations with the largest values of $|x_{i3}|$ and neglecting the rest, the sign of each equation being chosen such that the coefficient of β_3 becomes positive. The fourth combination is obtained by the same procedure applied to $|x_{i2}|$. In this way the matrix of coefficients in the resulting four equations becomes nearly diagonal which gives a good determination of the unknowns.

Laplace calculates the residuals, not only for the 24 observations but also for 19 further observations from the same period. He remarks that some of the residuals are larger than expected from a knowledge of the size of errors of observation and that the pattern of the signs shows that a small systematic variation still exists.

Stigler ([250] p.34) has analyzed the 24 observations by the method of least squares and tabulated the residuals, which do not deviate essentially from those found by Laplace.

Mayer's and Laplace's methods are special cases of the theory of linear estimation. However, neither Mayer nor Laplace gave an algebraic formulation or a general discussion of their methods; they only solved the problem at hand. Their procedure became widely used and existed for many years as a competitor to the method of least squares because it produced good results with much less numerical work.

6.3 The Method of Least Absolute Deviations by Boscovich, 1757, and Laplace, 1799

Roger Joseph Boscovich (1711–1787) was born in Dubrovnik where he attended a Jesuit school. As fifteen years old he was sent to Rome to complete his training as a Jesuit priest; he was ordained in 1744. He also studied mathematics, astronomy, and physics and published papers in these fields. He became professor of mathematics at the Collegium Romanum in 1740.

Between 1735 and 1754 the French Academy carried out four measurements of the length of an arc of a meridian at widely different latitudes with the purpose of determining the figure of the Earth, expressed as its ellipticity. Pope Benedict XIV wanted to contribute to this project and in 1750 he commissioned Boscovich and the English Jesuit Christopher Maire to measure an arc of the meridian near Rome and at the same time to construct a new map of the Papal States. Their report was published in 1755.

The relation between arc length and latitude for small arcs is approximately $y = \alpha + \beta x$, where y is the length of the arc and $x = \sin^2 L$, where L is the latitude of the midpoint of the arc. The ellipticity equals $\beta/3\alpha$. From the measured arcs the length of a one-degree arc is calculated and used as the observed value of y. Boscovich's problem was to estimate α and β from the five observations of (x, y).

In 1757 he published a summary of the 1755 report in which he proposed to solve the problem of reconciling inconsistent linear relations by the following method. Minimize the sum of the absolute values of the residuals under the restriction that the sum of the residuals equals zero; that is, minimize $\sum |y_i - a - bx_i|$ with respect to a and b under the restriction $\sum(y_i - a - bx_i) = 0$. Boscovich was the first to formulate a criterion for fitting a straight line to data based on the minimization of a function of the residuals. His formulation and solution are purely verbal, supported by a diagram that explains the method of minimization. We give an algebraic solution that follows his mode of reasoning.

Using the restriction $\bar{y} = a + b\bar{x}$ to eliminate a we get

$$S(b) = \sum_{i=1}^{n} |y_i - \bar{y} - b(x_i - \bar{x})|,$$

which has to be minimized with respect to b. Setting and $X_i = x_i - \bar{x}, Y_i = y_i - \bar{y}$, and $b_i = Y_i/X_i$ (the slope of the line connecting the ith observation with the center of gravity) we get

$$S(b) = \sum_{i=1}^{n} |X_i||b_i - b|,$$

where Boscovich orders the observations such that $b_1 > b_2 > \cdots > b_n$. Hence, $S(b)$ is a piecewise linear function of b with a slope depending on the position of b in relation to the b_is. For $b_j > b > b_{j+1}, j = 1, \ldots, n-1$, the slope equals

$$\frac{S(b_j) - S(b_{j+1})}{b_j - b_{j+1}} = \sum_{i=j+1}^{n} |X_i| - \sum_{i=1}^{j} |X_i| = \sum_{i=1}^{n} |X_i| - 2\sum_{i=1}^{j} |X_i|. \qquad (6.2)$$

The minimum of $S(b)$ is thus obtained for $b = b_k$, say, where k is determined from the inequality

$$\sum_{i=k+1}^{n} |X_i| - \sum_{i=1}^{k} |X_i| \leq 0 < \sum_{i=k}^{n} |X_i| - \sum_{i=1}^{k-1} |X_i|, \qquad (6.3)$$

or equivalently from

$$\sum_{i=1}^{k-1} |X_i| < \frac{1}{2}\sum_{i=1}^{n} |X_i| \leq \sum_{i=1}^{k} |X_i|, \qquad (6.4)$$

which is the form used by Boscovich. Today b_k is called the weighted median of the b_is.

In a discussion of the figure of the Earth, Laplace [153] proves Boscovich's result simply by differentiation of $S(b)$. Supposing that $b_k > b > b_{k+1}$ he gets

$$S'(b) = -\sum_{i=1}^{k} |X_i| + \sum_{i=k+1}^{n} |X_i|,$$

which for $S'(b) \leq 0, b < b_k$, and $S'(b) > 0, b > b_k$, gives Boscovich's result (6.4).

In the *Mécanique Céleste*, ([154] Vol. 2) Laplace returns to the problem and proposes to use Boscovich's two conditions directly on the measurements of the arcs instead of the arc lengths per degree; that is, instead of y_i he considers $w_i y_i$, where w_i is the number of degrees. Hence, he minimizes $\sum w_i |y_i - a - b x_i|$ under the restriction $\sum w_i(y_i - a - b x_i) = 0$. Introducing X_i and Y_i as the deviations from the weighted means and setting $b_i = Y_i/X_i$ the problem is formally the same as above, so the value of k is found from (6.4) by substituting $w_i|X_i|$ for $|X_i|$.

Bowditch ([20] Vol. 2, p.438) points out that the method of least absolute deviations is preferable to the method of least squares for estimating the slope of the line if extreme errors occur.

The method of least absolute deviations had drawbacks compared with the method of averages and the method of least squares: (1) the estimate of the slope is nonlinear and complicated to calculate, and (2) the method was restricted to one independent variable. The method therefore disappeared from statistical practice until the second half of the twentieth century when questions of robustness of estimates were discussed.

6.4 The Method of Least Squares by Legendre, 1805

Adrien-Marie Legendre (1752–1833) got his basic education in mathematics and the natural sciences in Paris. He was professor of mathematics at the École Militaire in Paris from 1775 to 1780. The Academy of Sciences appointed him as a member of important committees on astronomical and geodetical projects, among them the committee for determining the standard meter based on measurements of a meridian arc through Paris. His main scientific work was concentrated on celestial mechanics, number theory and the theory of elliptic integrals. In competition with Laplace he worked on attraction and the figure of the Earth. He wrote a textbook on geometry and a treatise on the theory of elliptic functions in three volumes 1825–1828.

Legendre's "Nouvelle methods pour la determination des orbites des comètes" [165] contains an appendix (pp. 72–80) entitled "Sur la méthode des moindres carrés" in which for the first time the method of least squares is described as an algebraic procedure for fitting linear equations to data.

He begins by stating the linear measurement error model

$$y_i = b_1 x_{i1} + \cdots + b_m x_{im} + e_i, \; m \le n, \; i = 1, \ldots, n.$$

After some introductory remarks to the effect that one should proceed so that the extreme errors without regard to sign are contained within as narrow limits are possible, Legendre writes:

> Among all the principles that can be proposed for this purpose, I think there is no one more general, more exact, and more easy to apply than that which we have made use of in the preceding researches, and which consists in making the sum of the squares of errors a *minimum*. In this way there is established a sort of equilibrium among the errors, which prevents the extremes to prevail and is well suited to make us know the state of the system most near to the truth.

The sum of the squares of the errors is

$$\sum_{i=1}^{n} e_i^2 = \sum_{i=1}^{n} (y_i - b_1 x_{i1} - \cdots - b_m x_{im})^2.$$

To find its minimum, Legendre sets the derivative with respect to b_k, $k = 1, \ldots, m$, equal to zero, which leads to the m linear equations

$$\sum_{i=1}^{n} y_i x_{ik} = b_1 \sum_{i=1}^{n} x_{i1} x_{ik} + \cdots + b_m \sum_{i=1}^{n} x_{im} x_{ik}, \ \ k = 1, \ldots, m,$$

later called "the normal equations." Legendre remarks that they have to be solved by the ordinary methods, which presumably means by successive elimination of the unknowns.

He states that if some of the resulting errors are judged to be too large, then the corresponding equations should be discarded as coming from faulty observations. Finally, he notes that the arithmetic mean is obtained as a special case by minimizing $\sum(y_i - b)^2$, and that the center of gravity for the observed coordinates (y_{1i}, y_{2i}, y_{3i}), $i = 1, \ldots, n$, is obtained by minimization of

$$\sum [(y_{1i} - b_1)^2 + (y_{2i} - b_2)^2 + (y_{3i} - b_3)^2].$$

Legendre's exposition of the method of least squares is clear and concise. However, one may wonder why he did not discuss the new method in relation to Laplace's two methods based on the least value of the absolute errors (residuals).

Legendre demonstrates the new method by analyzing the same data as Laplace, namely the five measurements of the meridian arcs making up the total of the arc through Paris, and used for determining the standard meter. His result does not deviate essentially from that found by Laplace.

The importance of Legendre's method of least squares was recognized immediately by leading astronomers and geodesists in France and Germany.

7

Gauss's Derivation of the Normal Distribution and the Method of Least Squares, 1809

7.1 Biography of Gauss

Carl Friedrich Gauss (1777–1855) was born into a humble family in Brunswick, Germany. His extraordinary talents were noted at an early age, and his father allowed him to enter the local Gymnasium in 1788, where he excelled in mathematics and numerical calculations as well as in languages. Impressed by his achievements, a professor at the local Collegium Carolinum recommended him to the Duke of Brunswick, who gave Gauss a stipend, which made it possible for him to concentrate on study and research from 1792 until 1806, when the Duke died. For three years Gauss studied mathematics and classics at the Collegium Carolinum; in 1795 he went to the University of Göttingen and continued his studies for another three years. From 1798 he worked on his own in Brunswick until he in 1807 became professor in astronomy and director of the observatory in Göttingen, where he remained for the rest of his life.

This academic career took place at a time of great political turmoil, first the French revolution, then the Napoleonic wars with the French occupation of Germany, and finally the liberal revolutions of 1830 and 1848. Nevertheless, Gauss succeeded in keeping up a steady scientific activity of great originality in pure and applied mathematics.

In his doctoral dissertation in 1799 he proved the fundamental theorem of algebra and showed that a real polynomial can be written as a product of linear and quadratic factors with real coefficients. Another early mathematical masterpiece was the *Disquisitiones arithmeticae* (*Arithmetical investigations* [99],) which became of great importance for the development of number theory; here he proved the law of quadratic reciprocity, previously proved incompletely by Legendre. This work established his fame as a mathematician. Throughout his life he contributed to algebra, number theory, analysis, special functions, differential equations, differential geometry, non-Euclidean geometry, and numerical analysis.

In mathematical astronomy he achieved a similar early recognition by calculating the orbit of the new planet Ceres, which had been observed for a short period of time in the beginning of 1801 but then disappeared. At the end of the year it was located at a position very close to that predicted by Gauss. From then on Gauss calculated the orbits of several other planets and finally published his methods in the *Theoria motus corporum coelestium* (Theory of the motion of the heavenly bodies, [100], which contains his first exposition of the method of least squares, based on the assumptions that the observations are normally distributed and that the prior distribution of the location parameters is uniform.

Regarding his invention of the method of least squares Gauss ([100], §186) writes: "Our principle, which we have made use of since the year 1795, has lately been published by Legendre .... " This statement naturally angered Legendre who responded with a letter, 1809, pointing out that "There is no discovery that one cannot claim for oneself by saying that one had found the same thing some years previously; but if one does not supply the evidence by citing the place where one has published it, this assertion becomes pointless and serves only to do a disservice to the true author of the discovery." In 1811 Laplace brought the matter of priority before Gauss, who answered that "I have used the method of least squares since the year 1795 and I find in my papers, that the month of June 1798 is the time when I reconciled it with the principle of the calculus of probabilities." In the TAP ([159] II, §24) Laplace writes that Legendre was the first to publish the method, but that we owe to Gauss the justice to observe that he had the same idea several years before, that he had used it regularly, and that he had communicated it to several astronomers, see Plackett [219] for a full discussion of the priority dispute.

As a pure mathematician Gauss worked alone; he did not have the patience to explain and discuss his ideas with other mathematicians. He kept a mathematical diary in which he noted his results but did not publish before the proofs were in perfect form. In applied mathematics he worked together with astronomers, geodesists; and physicists. Besides giving significant contributions to the mathematical, numerical, and statistical analyses of data, he carried out himself a large number of observations, measurements, and experiments. In particular, he took part in the triangulation of Hanover, beginning in 1818 and continuing the fieldwork during the summer months for eight years. The analysis of these data led him to his second version of the method of least squares ([103], [104]) based on the minimization of the expected loss, expressed as the squared error of estimation.

Gauss was much influenced by Laplace. His first proof of the method of least squares is based on inverse probability inspired by Laplace's 1774 paper. After having proved the central limit theorem, Laplace ([155],[156], [159]) turned to the frequentist view of probability and Gauss followed suit in his second proof.

Gauss's books, papers, and some of his letters have been published in 12 volumes in *Werke* (1863–1933).

7.2 Gauss's Derivation of the Normal Distribution, 1809

As explained in Section 5.5, Laplace [148] had formulated the principle for parametric statistical inference as follows. Specify the mathematical form of the probability density for the observations, depending on a finite number of unknown parameters, and define a method of estimation that minimizes the error of estimation. He had hoped in this way to show that the arithmetic mean is the best estimate of the location parameter in the error distribution but failed to do so because he used the absolute value of the deviation from the true value as the error of estimation, which led to the posterior median as estimate. The gap between statistical practice and statistical theory thus still existed when Gauss took over.

Gauss [100] solved the problem of the arithmetic mean by changing both the probability density and the method of estimation. He turned the problem around by asking the question: What form should the density have and what method of estimation should be used to get the arithmetic mean as estimate of the location parameter? He ([100], §177) writes:

> It has been customary to regard as an axiom the hypothesis that if any quantity has been determined by several direct observations, made under the same circumstances and with equal care, the arithmetic mean of the observed values gives the most probable value, if not rigorously, yet very nearly, so that it is always most safe to hold on to it.

Let $f(x-\theta)$ be the probability density of the observations and assume that $f(.)$ is differentiable and tends to zero for the absolute value of the error tending to infinity. It then follows from Laplace's principle of inverse probability that the posterior density of θ equals

$$p(\theta|\underline{x}) = f(x_1 - \theta) \cdots f(x_n - \theta) / \int f(x_1 - \theta) \cdots f(x_n - \theta)d\theta.$$

According to the quotation above Gauss requires that the most probable value, the mode of $p(\theta|\underline{x})$, should be set equal to the arithmetic mean $\bar{x}$. Hence, he gets the differential equation

$$\frac{\partial \ln p(\theta|\underline{x})}{\partial \theta} = 0, \text{ for } \theta = \bar{x} \text{ and all values of } n \text{ and } \underline{x}.$$

The solution is the normal distribution

$$p(x|\theta, h) = \frac{h}{\sqrt{\pi}} \exp[-h^2(x - \theta)^2], \quad -\infty < x < \infty, \quad -\infty < \theta < \infty, \quad 0 < h < \infty.$$
$$(7.1)$$

Laplace's exponent $m|x-\theta|$ is thus replaced by $[h(x-\theta)]^2$. As does Laplace, Gauss parameterizes the distribution by the inverse scale parameter.

Gauss's invention of the normal distribution marks the beginning of a new era in statistics. Natural scientists now had a two-parameter distribution which (1) led to the arithmetic mean as estimate of the true value and thus to a probabilistic justification of the method of least squares, (2) had an easily understandable interpretation of the parameter h in terms of the precision of the measurement method, and (3) gave a good fit to empirical distributions of observations, as shown by Bessel [10].

Assuming that h is known it follows from the principle of inverse probability that

$$p(\theta|\underline{x}) \propto \exp[-h^2(x - \theta)^2],$$

so the posterior mode is found by minimizing $\sum(x_i - \theta)^2$, which of course leads to the arithmetic mean. Because

$$\sum(x_i - \theta)^2 = \sum(x_i - \bar{x})^2 + n(\bar{x} - \theta)^2,$$

θ is normally distributed with mean $\bar{x}$ and precision $h\sqrt{n}$.

Gauss's probabilistic justification of the method of least squares thus rests on the assumptions that the observations are normally distributed and that the prior distribution of the location parameter is uniform.

For observations with different precisions Gauss minimizes $\sum h_i^2(x_i - \theta)^2$ which leads to the weighted mean $\bar{x} = \sum h_i^2 x_i / \sum h_i^2$.

Regarding the constant term, Gauss refers to "the elegant theorem first discovered by Laplace," which shows that he had read Laplace [148].

In modern notation we have $h = 1/\sigma\sqrt{2}$. In many contexts it is, however, convenient to use Gauss's notation.

Laplace's result [150] that for large n, $p(\theta|\underline{x})$ is approximately normal for an arbitrary density $f(x|\theta)$ is thus supplemented by Gauss's result that, for any n, $p(\theta|\underline{x})$ is normal if $f(x|\theta)$ is normal.

It is important to distinguish between the Gaussian method of using the posterior mode as estimate and the method of maximum likelihood. The two methods lead to the same estimate but are based on fundamentally different concepts. There has been some confusion on this matter in the literature.

It is a surprising fact that nobody before Bessel [10] studied the form of empirical error distributions, based on the many astronomical observations at hand. Presumably, they would then have realized that the error distributions on which they spent so much mathematical labor (the rectangular, triangular, and quadratic) were poor representatives of the real world.

7.3 Gauss's First Proof of the Method of Least Squares, 1809

Gauss generalizes his results for one unknown parameter to the linear normal model $y = X\beta + \varepsilon$ with n observations and m parameters, $m < n$. He assumes that the m vectors of X are linearly independent, that the βs are independently and uniformly distributed on the real line, and that the εs are independently and normally distributed with zero mean and known precision h.

Using the principle of inverse probability he gets

$$p(\beta|y) \propto \exp[-h^2(y - X\beta)'(y - X\beta)].$$

The posterior mode, b say, is obtained by minimizing $(y - X\beta)'(y - X\beta)$, which leads to the normal equations $X'Xb = X'y$. Gauss ([100], [101]) solves these equations by successive elimination of the unknowns, in this way obtaining an upper triangular system of equations that is solved by backward substitution. The reduced normal equations may be written as

$$Ub = GX'y, \ GX'X = U,$$

where U is an upper triangular matrix, and G is a lower triangular matrix with diagonal elements equal to unity.

To find the posterior distribution of β_m, Gauss integrates out the first $m-1$ variables of $p(\beta|y)$. To carry out the integration he transforms the quadratic form in the exponent into a weighted sum of squares using the matrices from the reduced normal equations. Because

$$(y - X\beta)'(y - X\beta) = (y - Xb)'(y - Xb) + (b - \beta)'X'X(b - \beta), \qquad (7.2)$$

where the first term on the right is independent of β, he introduces the new variables $v = U(\beta - b)$ and proves that the second term on the right equals $v'D^{-1}v$, where the elements of the diagonal matrix D are the diagonal elements of U. Hence,

$$p(\beta|y) \propto \exp(-h^2 v' D^{-1} v). \qquad (7.3)$$

Integrating successively with respect to $v_1, \ldots, v_{m-1}$ and using that

$$v_m = u_{mm}(\beta_m - b_m),$$

Gauss finds

$$p(\beta_m|y) = \pi^{-1/2} h u_{mm}^{1/2} \exp[-h^2 u_{mm}(\beta_m - b_m)^2], \qquad (7.4)$$

so the posterior mode equals b_m, and β_m is normally distributed with mean b_m and squared precision $h^2 u_{mm}$.

To find the marginal distribution of $\beta_r, r = 1, \ldots, m - 1$, Gauss uses the fundamental equation

$$X'y = X'X\beta - z = X'Xb, \ z = -X'\varepsilon, \qquad (7.5)$$

which multiplied by G gives

$$v = U(\beta - b) = Gz.$$

In particular,

$$z_m = c_{m1}v_1 + \cdots + c_{m,m-1}v_{m-1} + v_m, \tag{7.6}$$

say, because G is unit lower triangular. Gauss then considers another form of the solution of the normal equations by introducing a matrix Q defined as $QX'X = I_m$. Multiplying (7.5) by Q he gets $\beta - b = Qz$. Hence

$$\beta_m - b_m = q_{m1}z_1 + \cdots + q_{mm}z_m = v_m u_{mm}^{-1}.$$

Using (7.6) it follows that $q_{mm} = u_{mm}^{-1}$, so that the squared precision of β_m is h^2/q_{mm}. From the equation

$$\beta_r - b_r = q_{r1}z_1 + \cdots + q_{rm}z_m,$$

it then follows by symmetry that the squared precision of β_r is h^2/q_{rr}.

Gauss does not discuss the covariances, neither does he discuss how to estimate h.

The proof demonstrates Gauss's mastery of linear algebra. His algorithm for inverting the symmetric matrix $X'X$ to obtain $b = (X'X)^{-1}X'y$ became a standard method in numerical analysis.

In his proofs Gauss uses a simple notation for the inner product of two vectors, a and b say, setting $\sum a_i b_i = [ab]$. We use this symbol in the following.

7.4 Laplace's Large-Sample Justification of the Method of Least Squares, 1810

Laplace had the main idea that the soundness of a statistical method should be judged by its performance in large samples. Having just proved the central limit theorem he [155],[156] immediately used it in his comments on Gauss's result.

For an arbitrary symmetric distribution with location parameter θ and finite variance he notes that $\bar{x}$ is asymptotically normal with mean θ and precision $h\sqrt{n}$, and combining k samples from the same population he gets

$$p(\theta|\bar{x}_1,\ldots,\bar{x}_k) \propto \exp[-\sum h_i^2 n_i(\bar{x}_i - \theta)^2]$$

for large samples, according to the principle of inverse probability. The posterior median equals the posterior mode, and the common value is the weighted mean

$$\bar{x} = \sum h_i^2 n_i \bar{x}_i / \sum h_i^2 n_i,$$

which is the value obtained by minimizing $\sum h_i^2 n_i(\bar{x}_i - \theta)^2$. Hence, for large samples the method of least squares is valid under weaker assumptions than those used by Gauss. Laplace remarks that this is a reason for using the method also for small samples.

Laplace points out that among all differentiable, symmetric error distributions the normal is the only one leading to the arithmetic mean as the posterior mode.

Having thus presented a large-sample theory based on his own principles one would have expected him to use it in the following. However, we have here reached a turning point in Laplace's theory of estimation. He had just developed a new theory based on the central limit theorem leading to the sample distribution of the arithmetic mean and the corresponding frequency interpretation of the method of least squares. In choosing between the two methods Laplace ([159] II, §23) remarks that because "we are in complete ignorance of the law of error for the observations" we are unable to specify the equation from which the inverse probability estimate should be obtained. We should therefore keep to the method of least squares which does not require a specification of the distribution but only the existence of the second moment.

8

Credibility and Confidence Intervals by Laplace and Gauss

8.1 Large-Sample Credibility and Confidence Intervals for the Binomial Parameter by Laplace, 1785 and 1812

It follows from Laplace's 1774 and 1785 papers that the large-sample inverse probability limits for θ are given by the relation

$$P(h - u\sqrt{h(1-h)/n} < \theta < h + u\sqrt{h(1-h)/n}|h) \cong \Phi(u) - \Phi(-u), \quad (8.1)$$

for $u > 0$. In 1812 ([159], II, §16) he uses the normal approximation to the binomial to find large-sample direct probability limits for the relative frequency as

$$P(\theta - u\sqrt{\theta(1-\theta)/n} < h < \theta + u\sqrt{\theta(1-\theta)/n}|\theta) \cong \Phi(u) - \Phi(-u). \quad (8.2)$$

Noting that $\theta = h + O(n^{-1/2})$ so that

$$\sqrt{\theta(1-\theta)/n} = \sqrt{h(1-h)/n} + O(n^{-1})$$

and neglecting terms of the order of n^{-1} as in the two formulas above he solves the inequality (8.2) with respect to θ and obtains for $u > 0$

$$P(h - u\sqrt{h(1-h)/n} < \theta < h + u\sqrt{h(1-h)/n}|\theta) \cong \Phi(u) - \Phi(-u). \quad (8.3)$$

The limits for θ in (8.1) and (8.3) are the same but the probabilities have different interpretations as indicated by our use of the modern notation for conditional probabilities which did not exist at the time of Laplace. However, Laplace explains the distinction clearly by stating that (8.3) refers to the probability of events whereas (8.1) refers to the probability of causes. This important remark implies that Laplace's previous results for binomial variables, derived by inverse probability, may be interpreted in terms of direct probability.

Today the limits are called credibility and confidence limits, respectively.

8.2 Laplace's General Method for Constructing Large-Sample Credibility and Confidence Intervals, 1785 and 1812

From Laplace's 1785 limit theorem it follows that θ is asymptotically normal with mean $\hat{\theta}$ and variance $\sigma^2(\theta) = (-D^2 \ln p(\hat{\theta}))^{-1}$; see (5.27). Hence, the credibility interval for θ equals $\hat{\theta} \pm u\sigma(\theta)$ with credibility coefficient $P(u) \simeq \Phi(u) - \Phi(-u)$, $u > 0$. The interval in (8.1) is a special case.

Similarly, in [159], he uses the central limit theorem to generalize (8.3). He finds confidence intervals for the absolute moments and for a regression coefficient but does not formulate a general rule. His method may, however, be described as follows.

Let t_n be asymptotically normal with mean θ and variance σ^2/n so that the probability limits for t_n equal $\theta \pm u\sigma_n n^{-1/2}$ with covering probability $P(u)$. Solving for θ Laplace finds the limits $t_n \pm u\sigma_n n^{-1/2}$. He remarks that σ_n should be estimated from the sample with an accuracy of order $n^{-1/2}$, which gives the interval $t_n \pm us_n n^{-1/2}$. The estimate s_n is obtained by applications of the central limit theorem, as we show in the examples. He assumes that $P(u)$ is not changed essentially if n is large. He remarks with satisfaction that he has thus found "an expression in which everything is given by the observations."

Laplace's examples are based on the measurement error model with finite moments.

Let $m_{(r)} = \sum |\varepsilon_i|^r/n$ denote the rth absolute moment and $\mu_{(r)} = E(|\varepsilon^r|)$ the corresponding true value, $r = 1, 2, \ldots$. By means of the characteristic function Laplace ([159], II, §19) proves that $m_{(r)}$ is asymptotically normal with mean $\mu_{(r)}$ and variance $(\mu_{(2r)} - \mu_{(r)}^2)/n$, which also follows directly from the central limit theorem. Hence, the confidence interval for $\mu_{(r)}$ is

$$m_{(r)} \pm u(m_{(2r)} - m_{(r)}^2)^{1/2} n^{-1/2}. \tag{8.4}$$

For the regression model $x_i = \beta z_i + \varepsilon_i$ he proves that, within the class of linear estimates of β, the least squares estimate $b = [zx]/[zz]$ minimizes the absolute value of the error of estimation and that b is asymptotically normal with mean β and variance $\sigma^2/[zz]$. He ([159], II, pp. 326–327) then finds the confidence limits for β as $b \pm us[zz]^{-1/2}$, where $s^2 = \sum(x_i - bz_i)^2/n$.

8.3 Credibility Intervals for the Parameters of the Linear Normal Model by Gauss, 1809 and 1816

Assuming that h is known, Gauss [100] finds limits for the regression coefficients, and assuming that the true value of the variable in question is known, he [102] finds limits for the precision h. He does not discuss the case where both quantities are unknown.

As shown in Section 7.3, Gauss [100] proves that the squared precision of β_r is h^2/q_{rr}, where $Q = (X'X)^{-1}$. This means that $V(\beta_r) = \sigma^2 q_{rr}$, so the credibility interval for β_r equals $b_r \pm u\sigma q_{rr}^{1/2}$.

Let $\underline{\varepsilon} = (\varepsilon_1, \ldots, \varepsilon_n)$ be independently and normally distributed variables with mean zero and precision h, and let h be uniformly distributed on $[0, \infty)$ so

$$p(h|\underline{\varepsilon}) \propto h^n \exp(-h^2[\varepsilon\varepsilon]).$$

Gauss concludes that the most probable value of h is $\hat{h} = \sqrt{n/2[\varepsilon\varepsilon]}$, and because $\sigma = 1/h\sqrt{2}$ he sets $\hat{\sigma} = \sqrt{[\varepsilon\varepsilon]/n}$. Expanding $\ln p(\hat{h} + x)$ in Taylor's series he shows that h is asymptotically normal with mean $\hat{h}$ and variance $\hat{h}^2/2n$. The credibility interval for h is thus $\hat{h}(1 \pm u/\sqrt{2n})$ from which he by substitution finds the interval for σ as $\hat{\sigma}(1 \pm u/\sqrt{2n})$.

8.4 Gauss's Rule for Transformation of Estimates and Its Implication for the Principle of Inverse Probability, 1816

Gauss's brief paper [102] "Bestimmung der Genauigkeit der Beobachtungen" represents a turning point in his approach to estimation theory. First, he ends his applications of inverse probability by finding credibility limits as discussed in Section 8.3. Next, he refers to Laplace and the central limit theorem indicating that in the future he will use probability in the frequency sense. Finally, he derives the asymptotic relative efficiency of some estimates of the standard deviation in the normal distribution; see Section 8.5.

We here discuss the implications of his estimation of the precision and the standard deviation by inverse probability. After having found the most probable value of h as $\hat{h} = \sqrt{n/2[\varepsilon\varepsilon]}$ he writes (in our notation):

> The most probable value of σ is consequently $1/h\sqrt{2}$. This result holds generally, whether n be large or small.

Hence, Gauss transforms $\hat{h}$ to $\hat{\sigma} = \sqrt{[\varepsilon\varepsilon]/n}$ as if the estimates were parameters.

The general form of this rule is as follows. If t_n is the most probable value of θ, then $g(t_n)$ is the most probable value of $g(\theta)$ for any one-to-one transformation and all n. This rule was accepted by most statisticians, although it is clear that it violates the principle of inverse probability.

Expressed in terms of prior distributions, it can be said that Gauss used the principle of inverse probability as follows: to estimate h he assumed that the prior distribution of h is uniform, and to estimate a one-to-one transformation of h, $\sigma = \sigma(h)$ say, he assumed the prior distribution of σ to be uniform. However, this is equivalent to using the method of maximum likelihood. He thus realized that the posterior mode is not invariant to parameter transformations.

8.5 Gauss's Shortest Confidence Interval for the Standard Deviation of the Normal Distribution, 1816

Gauss uses Laplace's result for the absolute moments under the assumption that the εs are normally distributed $(0, \sigma^2)$. It follows that

$$\mu_{(r)} = a_r \sigma^r, \ a_r = \pi^{-1/2} 2^{r/2} \Gamma((r+1)/2), \ r = 0, 1, \ldots,$$

so that $\sigma = [\mu_{(r)}/a_r]^{1/r}$ with the corresponding estimate $s_r = [m_{(r)}/a_r]^{1/r}$. Among these estimates he proposes to find the one leading to the shortest confidence interval for σ.

Taking the rth root of Laplace's probability limits for $m_{(r)}$, he gets the limits for s_r as

$$\sigma(1 \pm ub_r n^{-1/2}), \ b_r = (a_{2r} - a_r^2)/r^2 a_r^2.$$

Solving for σ he obtains the confidence interval $s_r(1 \pm ub_r n^{-1/2})$. Tabulating b_r he shows that the shortest interval is obtained for $r = 2$, which gives

$$s_2(1 \pm u/\sqrt{2n}), \ s_2 = [\varepsilon\varepsilon]/n.$$

He remarks that "One hundred errors of observation treated by the formula for $r = 2$ will give a result as reliable as 114 treated by the formula for $r = 1$," and so on.

Finally, he notes that the confidence interval above equals the credibility interval previously found; see Section 8.3.

9

The Multivariate Posterior Distribution

9.1 Bienaymé's Distribution of a Linear Combination of the Variables, 1838

Irénée Jules Bienaymé (1796–1878) proposes to generalize Laplace's inverse probability analysis of the binomial. Using the principle of inverse probability on the multinomial he gets the posterior distribution

$$p_n(\theta_1, \ldots, \theta_k | n_1, \ldots, n_k) \propto \theta_1^{n_1} \cdots \theta_k^{n_k}, \ 0 < \theta_i < 1, \ \sum \theta_i = 1, \quad (9.1)$$

where the ns are nonnegative integers and $\sum n_i = n$. In normed form this distribution is today called the Dirichlet distribution. The posterior mode is $h_i = n_i/n, \sum h_i = 1$.

Bienaymé considers the linear combination $z = \sum c_i \theta_i$ and, assuming that $h_i > 0$ is kept constant as $n \to \infty$, he proves that z is asymptotically normal with mean $\bar{c} = \sum c_i h_i$ and variance $= \sum (c_i - \bar{c})^2 h_i / n$. There are no new principles involved in his proof but it is lengthy and complicated because z is a function of $k - 1$ correlated random variables. The proof has been discussed by von Mises [177], [178], [179] and by Heyde and Seneta [123].

We, however, give a simpler proof by using Lagrange's proof, [146], of the asymptotic normality of the Dirichlet distribution [see (3.16)], which was overlooked by Bienaymé and later authors. Because of the normality we need only find the limiting value of the first two moments. It is easy to prove that $E(\theta_i) \to h_i, nV(\theta_i) \to h_i(1 - h_i)$, and $nCV(\theta_i, \theta_j) \to -h_i h_j, i \neq j$, so that $E(z) \to \sum c_i h_i = \bar{c}$ and

$$nV(z) \to \sum c_i^2 h_i (1 - h_i) - \sum_{i \neq j} c_i c_j h_i h_j = \sum (c_i - \bar{c})^2 h_i.$$

9.2 Pearson and Filon's Derivation of the Multivariate Posterior Distribution, 1898

Pearson and Filon attempt to construct a general large-sample theory of estimation by starting from a multivariate distribution with density

$$f(x_1, \ldots, x_m | \theta_1, \ldots, \theta_k) \tag{9.2}$$

assuming that the parameters are uniformly distributed so that

$$p(\theta_1, \ldots, \theta_k | S) \propto \prod_{i=1}^{n} f(x_{i1}, \ldots, x_{im} | \theta_1, \ldots, \theta_k),$$

where S denotes the sample values indicated on the right side. Expanding this function in Taylor's series around the posterior mode $(\hat{\theta}_1, \ldots, \hat{\theta}_k)$, and neglecting terms of the third and higher degree in the deviations, they find that the distribution of $(\hat{\theta}_1 - \theta_1, \ldots, \hat{\theta}_k - \theta_k)$ is asymptotically normal with zero mean and inverse dispersion matrix $i(\underline{\theta}) = \{i_{rs}(\underline{\theta})\}$, where $\underline{\theta} = (\theta_1, \ldots, \theta_k)$, $\underline{x} = (x_1, \cdots, x_m)$, and $(r, s) = 1, \ldots, k$.

$$i_{rs}(\underline{\theta}) = -n \int \cdots \int \left(\frac{\partial^2 \ln f(\underline{x}|\underline{\theta})}{\partial \theta_r \partial \theta_s} \right) f(\underline{x}|\underline{\theta}) dx_1 \cdots dx_m. \tag{9.3}$$

However, this paper is unsatisfactory in several respects. They do not state explicitly that the estimates have to satisfy the equations

$$\sum \frac{\partial \ln f(x_{i1}, \ldots, x_{im} | \hat{\theta}_1, \ldots, \hat{\theta}_k)}{\partial \hat{\theta}_r} = 0, \ r = 1, \ldots, k,$$

and the inverse dispersion matrix should have been $j(\underline{\hat{\theta}}) = \{j_{rs}(\underline{\hat{\theta}})\}$, where

$$j_{rs}(\underline{\hat{\theta}}) = \sum \frac{\partial^2 \ln f(x_{i1}, \ldots, x_{im} | \hat{\theta}_1, \ldots, \hat{\theta}_k)}{\partial \hat{\theta}_r \partial \hat{\theta}_s}, \ (r, s) = 1, \ldots, k; \tag{9.4}$$

that is, they do not distinguish clearly between sample and population values. With these corrections, their proof is a generalization of Laplace's univariate proof, see (5.31). In the applications they work out $i(\underline{\theta})$ for several univariate and multivariate distributions in common use. The applications reveal that they have not understood that the theorem holds only for the posterior mode, not for estimates derived by the method of moments, unless the two methods lead to the same estimates, as pointed out by Fisher [67].

Edgeworth [47] gives the correct interpretation of the theorem.

The matrices $i(\underline{\theta})$ and $j(\underline{\theta})$ are called the expected and the observed information matrix, respectively.

Edgeworth's Genuine Inverse Method and the Equivalence of Inverse and Direct Probability in Large Samples, 1908 and 1909

10.1 Biography of Edgeworth

Francis Ysidro Edgeworth (1845–1926) was a complex personality with wide-ranging interests in both the humanities and the natural sciences. For several years he studied the classics at the universities of Dublin and Oxford, next he studied commercial law and qualified as a barrister, and finally he studied logic and mathematics on his own, using the acquired knowledge to write important books on ethics, utility, and economics.

In 1880 he became lecturer in logic at King's College, London, in 1888 he was promoted to professor of political economy, and in 1891 he was appointed professor in that topic at Oxford. Besides being one of the leading British economists, he wrote a large number of papers on probability and statistics. As did Laplace and Gauss, he wrote on inverse as well as direct probability, an important contribution is the Edgeworth series that generalizes Laplace's central limit theorem. Stigler [250] has discussed Edgeworth's work. Here we discuss some of his contributions to inverse probability.

10.2 The Derivation of the t Distribution by Lüroth, 1876, and Edgeworth, 1883

The estimation of the parameters of the normal distribution by inverse probability was treated by Gauss and his followers as two independent problems, in each case assuming one of the parameters as known. However, in a paper that has been overlooked, Lüroth [171] continued Gauss's analysis of the linear normal model by considering the joint distribution of the parameters, as pointed out by Pfanzagl and Sheynin [217]. Starting from

$$p(\beta, h|y) \propto h^n \exp[-h^2(y - X\beta)'(y - X\beta)], \qquad (10.1)$$

tacitly assuming that β and h are uniformly distributed, Lüroth derives the marginal distribution of β_m. Using the Gaussian transformation

$$(y - X\beta)'(y - X\beta) = [ee] + v'D^{-1}v,$$

and noting that

$$\int \exp(-h^2 z^2)dz \propto h^{-1},$$

he finds

$$p(\beta_m, h|y) = \int p(\beta, h|y)d\beta_1 \cdots d\beta_{m-1} \propto h^{n-m+1}\exp(-h^2([ee] + u_{mm}^{-1}v_m^2)).$$

The coefficient of $-h^2$ in the exponent may be written as

$$[ee]\left(1 + \frac{u_{mm}(\beta_m - b_m)^2}{[ee]}\right) = [ee]\left(1 + \frac{t^2}{(n - m + 1)}\right),$$

where

$$t^2 = \frac{u_{mm}(\beta_m - b_m)^2}{[ee]/(n - m + 1)}. \tag{10.2}$$

Integration with respect to h then gives

$$p(\beta_m|y) \propto \left(1 + \frac{t^2}{(n - m + 1)}\right)^{(n-m+2)/2}, \tag{10.3}$$

which is Lüroth's generalization of (7.4). It is shown that Lüroth's result is the t distribution with $n - m + 1$ degrees of freedom.

Lüroth states that the (inverse probability) interval for β_m derived from (10.3) holds "regardless of which value h may have," in contradistinction to (7.4) which supposes that h is known. He says that it is customary to replace σ in (7.4) by $s = \{[ee]/(n - m)\}^{1/2}$.

He compares the length of the corresponding intervals for a covering probability of 50 percent and concludes that there is no essential difference, so one can safely use the old and simpler method based on the normal distribution. He thus overlooks the large effect of the t distribution for small samples and for large covering probabilities.

Edgeworth [43], who did not know Lüroth's paper, asks the fundamental question: How can one find limits for the mean θ of a normal distribution when σ is unknown? The same question was later asked by Gosset [254] in a frequentist setting. Edgeworth derives the marginal distribution of θ for a uniform distribution of h, which gives

$$p(\theta|\underline{x}) = \int p(\theta|h, \underline{x})dh \propto \{[ce] + n(\bar{x} - \theta)^2\}^{-(n+1)/2}.$$

Hence,

$$t = (\theta - \bar{x})\sqrt{n}/\sqrt{[ee]/n} \tag{10.4}$$

is distributed as Student's t with n degrees of freedom. He remarks that the 50 per cent credibility interval for θ, depending on a knowledge of σ, should be replaced by the interval

$$\bar{x} \pm t n^{-1/2} \sqrt{[ee]/n}, \ P(t) - P(-t) = 0.5, \tag{10.5}$$

where $P(t)$ denotes the distribution function for t.

Returning to the problem he [255] makes the same mistake as Lüroth by concluding that the effect of using the t distribution instead of the normal is insignificant, because he considers only intervals with a credibility of 50 percent.

10.3 Edgeworth's Genuine Inverse Method, 1908 and 1909

For n observations from a normal distribution with unknown mean θ and known variance σ^2 Laplace had pointed out that the sampling distribution of $\bar{x}$ and the posterior distribution of θ have the common density

$$\sqrt{n/2\pi}\sigma^{-1} \exp\{-n(\theta - \hat{\theta})^2/2\sigma^2\}, \ \hat{\theta} = \bar{x}.$$

Edgeworth [47] extends this dual interpretation, which he calls the "reversibility of the inverse and direct point of view" to the nonnormal case. For the density $f(x - \theta)$ the posterior mode is determined from the equation

$$\frac{\partial \ln p(\theta|\underline{x})}{\partial \theta} = \sum \frac{\partial \ln f(x_i - \theta)}{\partial \theta} = 0 \text{ for } \theta = \hat{\theta}.$$

The large-sample distribution of θ is normal $(\hat{\theta}, 1/j(\hat{\theta}))$, where

$$j(\theta) = -\sum_{i=1}^{n} \frac{\partial^2 \ln f(x_i|\theta)}{\partial \theta^2}. \tag{10.6}$$

This result explains the name given to $j(\theta)$; the information in the sample about θ equals the reciprocal of the variance in the distribution of θ.

The corresponding expectation is

$$i(\theta) = -n \int \frac{\partial^2 \ln f(x|\theta)}{\partial \theta^2} f(x|\theta)dx. \tag{10.7}$$

Because $\theta - \hat{\theta} = O(n^{-1/2})$, $j(\hat{\theta})$ may be approximated by $i(\theta)$, so that for large samples

$$(j(\hat{\theta})/2\pi)^{1/2}\exp\{-\frac{1}{2}(\theta - \hat{\theta})^2 j(\hat{\theta})\}$$

$$\simeq (i(\theta)/2\pi)^{1/2}\exp\{-\frac{1}{2}(\theta - \hat{\theta})^2 i(\theta)\}. \tag{10.8}$$

This formula represents the equivalence of inverse and direct probability in estimation theory. It is a generalization and an explicit formulation of

Laplace's asymptotic equivalence of the two modes of inference. From the right side of (10.8) Edgeworth concludes that the sampling distribution of $\hat{\theta}$ is asymptotically normal $(\theta, 1/i(\theta))$. Referring to Pearson and Filon [213], he generalizes (10.8) to the multivariate case.

In some examples for finite n he compares the posterior and the sampling distributions of estimates but does not reach a general result.

Edgeworth points out that the posterior mode is noninvariant to parameter transformations. However, limiting himself to large-sample theory he remarks that this fact is of no importance, because ordinary transformations are nearly linear in a neighborhood of $\hat{\theta}$ of order $n^{-1/2}$.

For large n he introduces the "genuine inverse method," which may be summarized as follows.

(1) Use a uniform distribution for the parameters in the model regardless of the parameterization chosen.

(2) Maximize the joint posterior distribution to find the estimates. This rule combined with the one above obviously leads to the maximum likelihood estimates.

(3) The distribution of the parameters is multivariate normal with the posterior mode as mean and the inverse of the observed information matrix as dispersion matrix.

(4) Interchanging the roles of parameters and estimates, it follows that the sampling distribution of the estimates is multivariate normal with the parameters as means and the inverse of the expected information matrix as dispersion matrix.

(5) The posterior mode minimizes the posterior expected squared error.

Edgeworth's imposing work thus completes the large-sample theory of statistical inference by inverse probability initiated by Laplace. Moreover, he establishes the equivalence of estimation theory based on posterior distributions and sampling distributions.

Criticisms of Inverse Probability

11.1 Laplace

Perhaps the strongest criticisms of inverse probability, although indirect, are implied by the fact that Laplace after 1811 and Gauss after 1816 based their theory of linear minimum variance estimation on direct probability. Nevertheless inverse probability continued to be used; only few of the critics went as far as to reject the theory completely.

Some doubt about the principle of indifference can be found in the early work of Laplace. Poisson (1837) is critical, and the principle is rejected by Cournot [33] as being arbitrary and subjective.

Laplace distinguishes between constant and variable causes. For the urn model the constant cause is the ratio of white to black balls, and the variable causes are the numerous circumstances in connection with the drawing of a ball that determine the outcome. Hence, when Laplace speaks of the probability of causes he means constant causes. Poisson ([222], §§ 27 and 63) points out that this usage differs from the one ordinarily used when discussing causality; in probability theory, he says, we consider a cause, relative to an event, as being the thing that determines the chance for the occurrence of the event; the cause can be a physical or a moral thing. Likewise, de Morgan ([187], p. 53) writes: "By a *cause*, is to be understood simply a state of things antecedent to the happening of an event, without the introduction of any notion of agency, physical or moral." We have followed de Morgan by using the term inverse probability for the topic which in French and German literature is called the probability of causes.

Laplace states that the assessment of probability depends partly on our knowledge and partly on our ignorance. In the case of perfect knowledge, which is unattainable for human beings, an event follows with certainty from its cause, an axiom known as the principle of sufficient reason. In the case of complete ignorance Laplace asserts that equal probabilities must be assigned to each of the possible alternatives, an axiom later called the principle of insufficient reason or indifference. However, in most cases some knowledge of the subject matter exists but Laplace does not succeed in formulating a rule for utilizing this information. We discuss some of his examples.

Assuming that the probability of heads is uniformly distributed on $[(1 - \alpha)/2, (1 + \alpha)/2]$, $0 < \alpha \le 1$, Laplace ([148], § 6) finds that the probability of getting n heads in succession equals

$$\frac{1}{\alpha} \int_{(1-\alpha)/2}^{(1+\alpha)/2} \theta^n d\theta = \frac{1}{\alpha(n+1)} \left(\frac{1}{2}\right)^{n+1} [(1+\alpha)^{n+1} - (1-\alpha)^{n+1}].$$

Setting $\alpha = 1$ he gets $1/(n+1)$, and for $\alpha \to 0$ the classical result $(\frac{1}{2})^n$ is obtained. Laplace remarks that the probability of getting two heads equals $\frac{1}{4} + \alpha^2/12$, which gives $\frac{1}{4} + 1/300$ if we are sure that the bias of the coin is at most 0.1. In the same paper he also uses a two-point prior symmetric about $\theta = 1/2$. Hence, in his first paper on inverse probability he tries out three different priors for the binomial parameter, two of them covering only a part of the complete parameter space. For an ordinary coin it is known that the probability of heads is near $\frac{1}{2}$, but it is difficult to give this knowledge a precise expression in the form of a prior distribution. Presumably for this reason, and also to avoid subjectivity and obtain mathematical simplicity Laplace uses only the uniform prior in his following papers, for example, in his analyses of the sex ratio at birth and of the death rate for a given age interval.

An example of the misuse of the urn model and inverse probability is the problem of the rising sun, whose history has been analysed by Zabell [275]. Hume had written that it is ridiculous to say that it is only probable for the sun to rise tomorrow. In an attempt to refute Hume, Price uses Bayes's theorem to derive the posterior probability

$$P(\theta > \frac{1}{2}|n) = (2^{n+1} - 1)/2^{n+1},$$

where θ is the probability of a sunrise and n denotes the number of sunrises observed. Buffon misunderstands this result by taking the probability to mean the probability for a sunrise tomorrow and by giving the odds as 2^n to 1. In the *Essai* ([160], p. XIII) Laplace corrects Buffon's errors and remarks that the conditional probability for a sunrise tomorrow after the occurrence of n risings equals $(n+1)/(n+2)$. However, as an astronomer he adds that the problem really is one of celestial mechanics and that at present nothing can stop the sun's course so far as one can see. Laplace's remarks imply (1) that the problem is one of "Newtonian induction," not of probabilistic induction, because the analogy with the urn model is false, and (2) if the problem nevertheless is treated probabilistically it should be treated correctly.

Laplace does not in this connection refer to his generalization of the rule of succession, which says that the conditional probability of getting m further successes after having got n successes equals $(n+1)/(n+m+1)$. Hence, if m is large relative to n this probability is small.

In his theory of estimation Laplace emphasizes that a point estimate should always be accompanied by its standard error so that it can be turned into an interval estimate. Similarly, in his large-sample theory of prediction he supplements the point predictor with its standard error. However, for small samples he uses the rule of succession without discussing the prediction interval. For example, he states blandly that having observed one success the probability of success in the next trial equals

$$P(S_2|S_1) = E(\theta|1) = 2/3.$$

Because the conditional density of θ in this case equals 2θ the most probable value of θ is unity, the median is $1/\sqrt{2}$, the average $2/3$, and $P(\theta > \alpha|1) = 1 - \alpha^{n+1}$, so the prediction interval becomes $0.1 \leq \theta \leq 1$ for a credibility of 0.99. Similarly we have $E(\theta|n) = (n+1)/(n+2)$ and $P(\theta > \alpha|n) = 1 - \alpha^{n+1}$, thus for $n = 10$, say, the point predictor is 0.92 and the prediction interval is $[0.66, 1]$.

The rule of succession is one of the most conspicuous and easily understandable results of Laplace's theory of inverse probability and it therefore became a target for philosophers and frequentists in their attempt to discredit the theory. However, the critics did not take the uncertainty of the prediction into account and they did not contrast the rule of succession with an alternative frequentist rule. It was not until much later that a frequentist rule was formulated based on the hypergeometric distribution which means that $n + m + 1$ in the denominator is replaced by $n + m$ showing that the information provided by the uniform prior counts for one further observation. Laplace's $(a+1)/(n+2)$ is thus replaced by $(a+1)/(n+1)$.

11.2 Poisson

Poisson ([222], § 32) extends Laplace's rule to a discrete prior. Consider an urn with N balls, some white and some black, and let the number of white balls, k say, be distributed uniformly on the integers from 1 to N. Let S_n denote the occurrence of n white balls in succession by drawings without replacement from the urn, $1 \leq n \leq N$. Noting that

$$\sum_{k=1}^{N} k^{(n)} = (N+1)^{(n+1)}/(n+1), \ k^{(n)} = k(k-1)\cdots(k-n+1),$$

we get the probability of a white ball in the next drawing as

$$\begin{aligned}
P(W_{n+1}|S_n) &= \sum P(W_{n+1}|k, S_n)p(k|S_n) \\
&= \sum \{(k-n)/(N-n)\}k^{(n)}/\sum k^{(n)} \\
&= (N-n)^{-1}\sum k^{(n+1)}/\sum k^{(n)} \\
&= (n+1)/(n+2).
\end{aligned}$$

Poisson carries out the proof only for $n = 1$ i.e., he finds $P(W_2|W_1) = 2/3$, but the generalization is straightforward. He remarks that the result is independent of N and thus equal to Laplace's rule. Poisson does not mention that a general proof, although more cumbersome, had been provided by Prevost and Lhuilier [225]. Zabell [275] has given an explanation of this remarkable result by means of exchangeability.

Expressing k^n as a linear combination of $k^{(n)}, k^{(n-1)}, \ldots, k^{(1)}$, it follows that for drawings with replacement we have

$$P(W_{n+1}|S_n) = \frac{n+1}{n+2}\{1 + O(N^{-1})\}.$$

Poisson finds

$$P(W_2|W_1) = \frac{2}{3}(1 + \frac{1}{2N}).$$

Having thus found $P(W_2|W_1)$ for a uniform prior corresponding to complete ignorance, Poisson remarks that knowledge of the process by which the urn is filled should be taken into regard. Suppose that we have a superpopulation, an urn containing N white and N black balls, and that two balls are chosen at random without replacement and put into another urn. From this subpopulation a ball is drawn, which turns out to be white. Poisson shows that the probability of the second ball being white equals

$$P(W_2|W_1) = \frac{N+1}{2N+1} = \frac{1}{2}(1 - \frac{1}{2N-1}),$$

which shows that a two-stage process with objective probabilities gives an essentially different result from the rule of succession. He presents several more complicated examples of this type.

11.3 Cournot

Cournot ([33], § 240) writes:

> Nothing is more important than to distinguish carefully between the double meaning of the term *probability*, sometimes taken in an objective sense and sometimes in a subjective sense, if one wants to avoid confusion and errors in the exposition of the theory as well as in the applications.

He lives up to this program by giving a clear exposition of the analysis of binomially distributed observations by direct and inverse probability, respectively, in his Chapter 8. He characterizes precisely the difference between the applications of Bayes's theorem in the case of an objective two-stage model and the subjective model based on the principle of indifference. He (§ 93) points out that the rule of succession leads to results disagreeing with ordinary betting behavior. Tossing a new coin and getting heads, nobody will bet two to one on getting heads in the next toss. Knowing that a woman at her first birth has borne a boy, nobody will bet two to one that the next birth will give a boy. He remarks that one should attempt to estimate the probability in question by observing the relative frequency of a male second birth among cases with a male first birth. Application of the rule of succession leads to a "futile and illusory conclusion." Cournot also distances himself from the applications of inverse probability to judicial decisions and evaluation of testimonies.

As shown by Laplace, the large-sample confidence and credibility limits for θ are the same. Referring to this result and to the fact that the confidence limits are independent of any hypothesis on θ, Cournot (§ 95) points out that it is only through this interpretation that the credibility limits "acquire an objective value." He thus rejects inverse probability and interprets credibility intervals as confidence intervals, but lacking an adequate terminology to distinguish between the two concepts he naturally speaks of the probability of θ even if he does not consider θ as a random variable.

In the same section he mentions that the inverse probability results hold whether the prior is uniform or not only if it is nearly constant in the neighborhood of the observed relative frequency.

For small samples he (§ 240) remarks that the results derived by inverse probability are illusory because they depend on subjective probabilities; they may be used for betting purposes but do not apply to natural phenomena.

11.4 Ellis, Boole, and Venn

The British empirical school of probability, beginning with R.L. Ellis [55] and J. Venn [263], proposes to define probability as the limit of the relative frequency of a certain attribute in an infinite series of independent trials or observations under the same essential circumstances; see Venn ([263], p. 163). Venn (p. 74) maintains that "Experience is our sole guide." From such a system the uniform prior based on the principle of indifference is obviously excluded. As another consequence the classical definition of probability as the ratio of the number of favorable cases to the total number of possible cases is abolished. Ellis considers Bernoulli's proof of the law of large numbers as superfluous; it seems to him to be true a priori.

Ellis and Venn fail to appreciate that Bernoulli's starting point is the same as their own, namely the empirical fact that "the more observations that are taken, the less the danger will be of deviating from the truth," (J. Bernoulli, [6], p. 225). Beginning with games of chance Bernoulli formulates the classical definition of probability which he then proceeds to use also in his discussion of problems where it is impossible to speak of equally possible cases; that is, he tacitly extends the definition to cover series with stable relative frequencies. To test the stability of a series of relative frequencies, for example, the yearly proportion of male births or the conviction rates in criminal trials, Poisson and Cournot use the standard error for binomial relative frequencies, thus providing an objective criterion for the applicability of the binomial model. Venn, however, presents a lengthy discussion of the characteristics that a series should possess for falling under his theory but does not indicate any objective method for reaching a conclusion.

By means of examples Ellis and Venn criticize Price's formula and Laplace's rule of succession and conclude that the results obtained by inverse probability are illusory.

G. Boole ([16], p. 363) writes about inverse probability that "It is, however, to be observed, that in all those problems the probabilities of the *causes* involved are supposed to be known *a priori*. In the absence of this assumed element of knowledge, it seems probable that arbitrary constant would *necessarily* appear in the final solution." Writing Bayes's formula as

$$P(C|E) = \frac{P(C)P(E|C)}{P(C)P(E|C) + P(\overline{C})P(E|\overline{C})},$$

where $\overline{C}$ denotes the complement of C, he (p. 367) concludes that the formula does not give a definite value of $P(C|E)$ unless there are means for determining the values of $P(C)$ and $P(E|\overline{C})$. "The equal distribution of our knowledge, or rather of our ignorance [...] is an arbitrary method of procedure" (p. 370). He points out that other constitutions of the system of balls in the urn than the one assumed by Laplace lead to results differing from the rule of succession and gives examples of such constitutions. Hence, "These results only illustrate the fact, that when the defect of data are supplied by hypothesis, the solution will, in general, vary with the nature of the hypotheses assumed" (p. 375). This conclusion had previously been reached by the actuary Lubbock [169], who points out that Laplace's assumption of a uniform prior for the probability of dying within a given age interval is at variance with experience. He derives a prediction formula with $d\theta$ replaced by $w(\theta)d\theta$ and proposes to use a polynomial density for θ.

The criticisms of inverse probability advanced by Ellis, Venn, and Boole did not add essential new points of view to that given by Cournot but it helped to disseminate the message. However, judging from the reaction of Jevons ([129], pp. 256–257) the arguments were not considered as decisive:

> It must be allowed that the hypothesis adopted by Laplace is in some degree arbitrary, so that there was some opening for the doubt which Boole has cast upon it (*Laws of Thought*, pp. 368–375). But it may be replied, (1) that the supposition of an infinite number of balls treated in the manner of Laplace is less arbitrary and more comprehensive than any other that can be suggested. (2) The result does not differ much from that which would be obtained on the hypothesis of any large finite number of balls. (3) The supposition leads to a series of simple formulas which can be applied with ease in many

cases, and which bear all the appearance of truth so far as it can be independently judged by a sound and practiced understanding.

11.5 Bing and von Kries

Two further arguments against inverse probability were provided by the Danish actuary Bing [15]. First, Bing points out that by drawing a random sample from an urn containing an unknown number of black and nonblack balls the posterior probability for the number of black balls depends, according to the indifference principle, on whether we consider the content of the urn as black and nonblack or as black, white, and yellow, say. Hence, the solution depends critically on whether the hypotheses considered can be subdivided into hypotheses of a similar nature.

Next, he considers an example by Laplace (TAP, II, § 30) on the posterior distribution of survival probabilities $p(\theta_1, \theta_2 | S)$, S denoting the sample and (θ_1, θ_2) the survival probabilities, independently and uniformly distributed on the unit interval a priori. Introducing the probabilities of dying,

$$\lambda_1 = 1 - \theta_1 \text{ and } \lambda_2 = \theta_1(1 - \theta_2),$$

Bing derives the density of (λ_1, λ_2) for given S by multiplying $p(\theta_1, \theta_2 | S)$ by the absolute value of the Jacobian of the transformation. He points out that if Laplace had started from the probabilities of dying and assumed a uniform distribution of (λ_1, λ_2) on the parameter space, which is the triangle $0 \leq \lambda_1 \leq 1$, $0 \leq \lambda_2 \leq 1$, $0 \leq \lambda_1 + \lambda_2 \leq 1$, then he would have found a different result. He derives both formulas. Hence, the posterior distribution obtained depends on the parameterization of the model, and he concludes that in cases where nothing can be said for preferring one set of parameters to another the indifference principle leads to contradictory results.

With hindsight we can see that Cournot and Bing between them produced all the arguments against inverse probability that have been advanced. However, their works were not widely read, and therefore the same arguments were later presented independently by many authors. We only mention a few.

The German logician von Kries [142] rediscovered Bing's two arguments. He presented the second in a much simpler form by noting that if $\theta > 0$ is uniform on a finite interval then $1/\theta$ is nonuniform on the corresponding interval. Because it is often arbitrary in which way a natural constant is measured, the principle of indifference leads to contradictory results; for example, should specific gravity or specific volume be considered as uniformly distributed. As did Cournot he observed that the shape of the prior distribution matters only in a small interval about the maximum of the likelihood function.

11.6 Edgeworth and Fisher

Edgeworth ([47], pp. 392 and 396) notes that if $h = 1/\sigma\sqrt{2}$ is uniformly distributed then the distribution of $c = 1/h$ is nonuniform, and that the same argument applies to the correlation coefficient and its square. He remarks, "There seems to be here an irreducible element of arbitrariness; comparable to the indeterminateness which baffles us when we try to define a 'random line' on a plane, or a 'random chord' of a circle." Nevertheless, he continues to use inverse probability, mainly because of the good asymptotic properties of the estimates.

Fisher [67] considers the binomial case with a uniform distribution of θ so that

$$p(\theta|a, n) \propto \theta^a(1 - \theta)^{n-a}, \ 0 < \theta < 1. \tag{11.1}$$

Making the transformation

$$\sin \lambda = 2\theta - 1, \ -\pi/2 \le \lambda \le \pi/2,$$

it follows that

$$p(\lambda|a, n) \propto (1 + \sin \lambda)^{a+1/2}(1 - \sin \lambda)^{n-a+1/2}, \tag{11.2}$$

because

$$d\lambda = \theta^{-1/2}(1 - \theta)^{-1/2}d\theta.$$

However, if λ had been assumed to be uniformly distributed the posterior density of λ is obtained from (11.1) by substituting $(1 + \sin \lambda)/2$ for θ, the result being inconsistent with (11.2).

In principle there is nothing new in Fisher's example but nevertheless it had a great effect, because it was included in his revolutionary 1922 paper where he introduced maximum likelihood estimation, which is invariant to parameter transformation. The peaceful coexistence of direct and inverse probability in Edgeworth's work was replaced by Fisher's maximum likelihood method and his aggressive ideological war against inverse probability.

More details on the history of the topics treated in the present chapter are given by Keynes [137], Zabell [275], [276], Dale [38], and Hald [114].

THE CENTRAL LIMIT THEOREM AND LINEAR MINIMUM VARIANCE ESTIMATION BY LAPLACE AND GAUSS

12

Laplace's Central Limit Theorem and Linear Minimum Variance Estimation

12.1 The Central Limit Theorem, 1810 and 1812

It is a remarkable fact that Laplace simultaneously worked on statistical inference by inverse probability, 1774–1786, and by direct probability, 1776–1781. In 1776 he derived the distribution of the arithmetic mean for continuous rectangularly distributed variables by repeated applications of the convolution formula. In his comprehensive 1781 paper he derived the distribution of the mean for independent variables having an arbitrary, piecewise continuous density. As a special case he found the distribution of the mean for variables with a polynomial density, thus covering the rectangular, triangular, and parabolic cases. In principle he had solved the problem but his formula did not lead to manageable results because the densities then discussed resulted in complicated mathematical expressions and cumbersome numerical work even for small samples. He had thus reached a dead end and it was not until 1810 that he returned to the problem, this time looking for an approximative solution, which he found by means of the central limit theorem.

Let $x_i, i = 1, 2, \ldots$, be a sequence of independent random variables with $E(x_i) = \mu_i$ and $V(x_i) = \sigma_i^2$. According to the central limit theorem the distribution of the sum $s_n = x_1 + \cdots + x_n$ converges to the normal distribution as $n \to \infty$ under certain conditions. For the versions considered by Laplace and Poisson the conditions are that the distribution of x_i has finite support and is nondegenerate. Hence, σ_i^2 is bounded away from zero and infinity, that is, there exist constants independent of i such that

$$0 < m < \sigma_i^2 < M < \infty \text{ for all } i. \tag{12.1}$$

It follows that $V(s_n)$ is of order n and

$$\max_{1 \le i \le n} \sigma_i^2 / \sum_{i=1}^{n} \sigma_i^2 \to 0, \text{ as } n \to \infty. \tag{12.2}$$

In the first version of the theorem, proved by Laplace ([155], [159]), it is assumed that the variables are identically distributed with expectation μ and variance σ^2. The distribution of s_n is then asymptotically normal $(n\mu, n\sigma^2)$ so that

$$u_n = \frac{s_n - n\mu}{\sigma\sqrt{n}} = \frac{(\bar{x}_n - \mu)\sqrt{n}}{\sigma} \tag{12.3}$$

is asymptotically normal $(0,1)$, where $\bar{x}_n = s_n/n$. It follows that

$$P\left(|\bar{x}_n - \mu| \leq \epsilon\right) \cong \Phi(\epsilon\sqrt{n}/\sigma) - \Phi\left(-\epsilon\sqrt{n}/\sigma\right) \to 1, \text{ as } n \to \infty. \tag{12.4}$$

If x_i is binomial $(1,p)$, $0 < p < 1$, then $\sigma^2 = p(1-p)$ and Bernoulli's law of large numbers and de Moivre's normal approximation to the binomial are obtained.

Laplace remarks that the theorem holds also for distributions with infinite support because the bounds for x_i enter the result through σ^2 only. As usual, his intuition was later proved right.

It is, of course, a remarkable result that s_n is asymptotically normal regardless of the distribution of the xs, if only σ^2 is bounded away from zero and infinity.

Not only the theorem but also the tool for proving it, namely the characteristic function and its inversion, are epoch-making. Laplace ([155], §§ 3 and 6; 1812, II, §§ 18 and 22) defines the characteristic function as

$$\psi(t) = E(e^{ixt}) = 1 + i\mu_1't - \mu_2't^2/2! + \cdots, \quad \mu_r' = E(x^r), \tag{12.5}$$

from which he gets

$$\ln \psi(t) = i\mu_1't - \sigma^2 t^2/2! + \cdots, \quad \sigma^2 = \mu_2' - \mu_1'^2. \tag{12.6}$$

Because the characteristic function for s_n equals $\psi^n(t)$, the expansion of its logarithm equals $n \ln \psi(t)$. The problem is to find the frequency function of s_n from the characteristic function.

Laplace assumes that x takes on the integer value k with probability p_k, $\sum p_k = 1$, so that

$$\psi(t) = \sum p_k \exp(ikt).$$

Using the fact that

$$\frac{1}{2\pi} \int_{-\pi}^{\pi} \exp(ikt)dt = \begin{cases} 0, & k \neq 0, \\ 1, & k = 0, \end{cases}$$

he gets the inversion formula

$$\frac{1}{2\pi} \int_{-\pi}^{\pi} \exp(-ikt)\psi(t)dt = p_k. \tag{12.7}$$

It follows that

$$P(s_n = n\mu'_1 + s) = \frac{1}{2\pi} \int_{-\pi}^{\pi} \exp(-ist - \tfrac{1}{2}n\sigma^2 t^2 + \cdots)dt$$

$$\sim \frac{1}{\sqrt{2\pi n\sigma}} \exp(-s^2/(2n\sigma^2)),$$

neglecting terms of smaller order of magnitude. Hence, $s_n - n\mu'_1$ is asymptotically normal $(0, n\sigma^2)$, so the arithmetic mean $\bar{x}$ is asymptotically normal $(\mu'_1, \sigma^2/n)$.

Because the characteristic function for wy, w being an arbitrary real number, equals $\psi(wt)$ Laplace finds that the linear combination $\sum w_j x_j$ with integers as coefficients is asymptotically normal with mean $\mu'_1 \sum w_j$ and variance $\sigma^2 \sum w_j^2$.

To prove the corresponding result for the continuous case, Laplace approximates the continuous density function by a discrete function with equidistant arguments. However, the limit process is unsatisfactory; an improved version was provided by Poisson [220].

Let $\varepsilon_1, \ldots, \varepsilon_n$ be independently and identically distributed errors with zero mean and variance σ^2 and consider two linear combinations $z_1 = [w_1\varepsilon]$ and $z_2 = [w_2\varepsilon]$. By means of the characteristic function

$$\psi(t_1, t_2) = E\{\exp(iz_1 t_1 + iz_2 t_2)\}$$

and the same method of proof as above, Laplace [157] shows that the asymptotic distribution of (z_1, z_2) is bivariate normal with zero means and covariance matrix $\sigma^2 W$, $W = \{[w_r w_s]\}, r = 1, 2$, so

$$p(z) = \frac{1}{2\pi\sigma^2 |W|^{1/2}} \exp(-z'W^{-1}z/2\sigma^2), \ z' = (z_1, z_2). \tag{12.8}$$

The generalization to the multivariate normal is obvious.

Laplace also finds the characteristic functions for the normal and the Cauchy distributions.

The central limit theorem and its use for constructing a large-sample theory of estimation for the parameters in the linear model is Laplace's second revolutionary contribution to probability theory and statistical inference. It is remarkable that he in 1774, 25 years old, created the theory of inverse probability, and that he, 61 years old, created a frequentist theory of estimation which he preferred to the previous one. The central limit theorem has ever since been the foundation for the large-sample theory of statistical inference.

12.2 Linear Minimum Variance Estimation, 1811 and 1812

Laplace found Gauss's probabilistic justification of the method of least squares unsatisfactory because it assumes that the observations are normally distributed and uses the posterior mode as the estimate because the method of least squares then follows. Laplace maintains that the best estimate of the location parameter is the one minimizing the expected estimating error $E(|\hat{\theta}-\theta|)$ for all θ. This is the same criterion as used in 1774 but now in a frequentist setting. Moreover he considers only linear combinations of observations as estimates because it is impractical to use other functions when the number of observations and the number of parameters are large. Because linear estimates are asymptotically normal and the expected estimating error then is proportional to the standard error the best estimate is the one having minimum variance.

For the simple linear model

$$y_i = \beta x_i + \varepsilon_i, \ i = 1, \ldots, n,$$

Laplace considers the linear combination

$$[wy] = \beta[wx] + [we].$$

Setting $\varepsilon_1 = \cdots = \varepsilon_n = 0$ and solving for β he obtains the estimate

$$\tilde{\beta} = [wy]/[wx],$$

which according to the central limit theorem is asymptotically normal with mean β and variance $\sigma^2[ww]/[wx]^2$, so

$$E(|\tilde{\beta} - \beta|) = \left(\frac{2}{\pi}\right)^{1/2} \sigma \frac{[ww]^{1/2}}{[wx]}.$$

Setting the logarithmic derivative of this expression with respect to w_i equal to zero Laplace finds that $w_i = cx_i$ for all i. Hence, the best linear estimate is $b = [xy]/[xx]$, which is asymptotically normal with mean β and variance $\sigma^2/[xx]$. As a corollary he remarks that b may be obtained by minimizing the sum of the squared observational errors $\sum(y_i - \beta x_i)^2$ so that the method of least squares leads to the best large-sample estimate whatever the error distribution if only its moments are finite. He proposes to estimate σ^2 by $s^2 = \sum(y_i - bx_i)^2/n$ so that the confidence limits for β are $b \pm us[xx]^{-1/2}$. For $x_1 = \cdots = x_n = 1$, the best estimate of β is the arithmetic mean.

For the linear model with two parameters

$$y_i = \beta_1 x_{i1} + \beta_2 x_{i2} + \varepsilon_i, \ i = 1, \ldots, n,$$

Laplace ([157], § 8; 1812, II, § 21) introduces two linearly independent vectors of coefficients which give the estimating equations

$$[w_1 y] = [w_1 x_1]\beta_1 + [w_1 x_2]\beta_2 + [w_1 \varepsilon], \qquad (12.9)$$
$$[w_2 y] = [w_2 x_1]\beta_1 + [w_2 x_2]\beta_2 + [w_2 \varepsilon].$$

Setting $\varepsilon_1 = \cdots = \varepsilon_n = 0$ and solving for (β_1, β_2) Laplace obtains the class of linear estimates, $(\tilde{\beta}_1, \tilde{\beta}_2)$ say, within which he seeks the one with minimum variance. It is shown that he here, without discussion, as in the one-parametric case, introduces the (obvious) restriction that the estimate should equal the true value if the observations are without error, a restriction that implies unbiasedness. Gauss uses the same restriction, and Sprott [242] proposes to call it "error-consistency."

Let us denote the matrix of coefficients in (12.9) by A and the linear combinations of errors by $z' = (z_1, z_2)$. Eliminating $[w_1 y]$ and $[w_2 y]$ Laplace obtains the equation $A\beta + z = A\tilde{\beta}$ and proves that $p(\tilde{\beta}) = p(z|A)$. From (12.8) Laplace finds by the substitution $z = A(\tilde{\beta} - \beta)$ that

$$p(\tilde{\beta}) = \frac{|A|}{2\pi\sigma^2 |W|^{1/2}} \exp\{-(\tilde{\beta} - \beta)'A'W^{-1}A(\tilde{\beta} - \beta)/2\sigma^2\}, \qquad (12.10)$$

from which he obtains the marginal distribution

$$p(\tilde{\beta}_1) = \frac{|A|}{(2\pi\sigma^2 H)^{1/2}} \exp - |A|^2 (\tilde{\beta}_1 - \beta_1)^2 / 2\sigma^2 H,$$
$$H = a_{12}^2 w_{22} - 2a_{12}a_{22}w_{12} + a_{22}^2 w_{11},$$

and

$$E|\tilde{\beta}_1 - \beta_1| = \left(\frac{2}{\pi}\right)^{1/2} \sigma \frac{\sqrt{H}}{|A|}. \qquad (12.11)$$

To minimize the mean error of estimation with respect to w_1 and w_2, Laplace sets the logarithmic derivative of $\sqrt{H}/|A|$ equal to zero from which he finds that the optimum values of w_1 and w_2 are proportional to x_1 and x_2, respectively. Hence, $W = A$ and the distribution of the best estimate equals

$$p(b) = \frac{|A|^{1/2}}{2\pi\sigma^2} \exp(-(b - \beta)'A(b - \beta)/2\sigma^2), \quad A = X'X. \qquad (12.12)$$

Replacing w_1 and w_2 by x_1 and x_2 the equations (12.9) become the normal equations, so the best estimate equals the least squares estimate.

Laplace remarks that this analysis can be extended to any number of parameters. He is mainly interested in the marginal distributions and gives a recursion formula for finding the variances, presenting explicit expressions for three and four parameters. Lacking an adequate notation his formula is complicated; in matrix notation it may be written as

$$V(b_r) = \sigma^2 A_{rr}/|A|, \quad r = 1, \ldots, m,$$

where A_{rr} denotes the cofactor of a_{rr}

It is clear that Laplace's asymptotic results hold for all n if the observations are normally distributed. However, Laplace does not discuss this case because he, when the distribution is known, would have used inverse probability, which would have led him to the solution given by Gauss [100] although from another point of view. Under normality direct and inverse probability lead to the same limits for β.

The method of least squares and the method of maximizing the posterior density are intuitively appealing. However, we owe to Laplace the fundamental observation that the justification of these methods depends on the properties of the estimates, that is, on the distribution of the error of estimation.

12.3 Asymptotic Relative Efficiency of Estimates, 1818

Laplace discusses asymptotic efficiency in the last section of the Second Supplement [162] to the TAP. For the linear model he remarks that whatever method is used for solving the equations leading to the class of linear estimates the result is that $\tilde{\beta}_1$, say, is expressed in the form

$$\tilde{\beta}_1 = [k_1 y] = [k_1 x_1]\beta_1 + \cdots + [k_1 x_m]\beta_m + [k_1 \varepsilon]. \qquad (12.13)$$

Choosing

$$[k_1 x_1] = 1, \ [k_1 x_2] = \cdots = [k_1 x_m] = 0, \qquad (12.14)$$

he obtains

$$\tilde{\beta}_1 = \beta_1 + [k_1 \varepsilon]$$

and

$$V(\tilde{\beta}_1) = \sigma^2 [k_1 k_1],$$

where k_1 depends on the ws and the xs, see (12.9). For $m = 2$ he finds

$$V(\tilde{\beta}_1) = \sigma^2 H / |A|^2,$$

which is a simple alternative proof of (12.10). The asymptotic efficiency of $\tilde{\beta}_1$ relative to b, the least squares estimate, is thus equal to the ratio of the corresponding values of $H/|A|^2$. Laplace derives the efficiency for two special cases in which the coefficients $k_1, \ldots, k_m$ are chosen such that the estimates are simpler to calculate than the least squares estimates.

Finally, Laplace compares his modification of Boscovich's nonlinear estimate with the best linear estimate for the model

$$y_i = \beta x_i + \varepsilon_i, \ x_i > 0, \ i = 1, \ldots, n,$$

assuming that the equations have been ordered such that

$$\frac{y_1}{x_1} > \frac{y_2}{x_2} > \cdots > \frac{y_n}{x_n}.$$

The estimate $\tilde{\beta}$, say, is defined as the value of β minimizing

$$\sum |y_i - \beta x_i| = \sum x_i \left| \frac{y_i}{x_i} - \beta \right|,$$

so $\tilde{\beta} = y_k/x_k$ where k is determined by the inequalities

$$\sum_1^{k-1} x_i < \sum_k^n x_i \text{ and } \sum_1^k x_i > \sum_{k+1}^n x_i.$$

Today this estimate is called the weighted median; the median is obtained for $x_1 = \cdots = x_n = 1$.

From the equation $v_k = \beta x_k + \varepsilon_k$ it follows that the error of estimation equals $v = \varepsilon_k/x_k$, which for convenience is supposed to be positive. Because $\varepsilon_j/x_j \lessgtr v$ for $j \lessgtr k$, Laplace finds the density of v as

$$p(v) \propto \prod_1^{k-1}[1 - F(x_j v)] \prod_{k+1}^n F(x_j v) f(x_k v),$$

where f denotes the density and F the distribution function of the εs. Assuming that f is symmetric about zero and using Taylor expansions of f, F, and $\ln p(v)$, Laplace proves that v is asymptotically normal with zero mean and variance equal to $1/(4f^2(0)[xx])$. Hence, the efficiency of the weighted median relative to the best linear estimate is $4f^2(0)\sigma^2$. Laplace concludes that the "method of situation" is preferable to the method of least squares if $[2f(0)]^2 > \sigma^{-2}$. He notes that for normally distributed observations the efficiency of the weighted median is $2/\pi$.

Next, Laplace takes the remarkable step of investigating the joint distribution of b and $\tilde{\beta}$ to find out whether a linear combination of the two will give a better estimate than b. Setting

$$z = \sum_1^{k-1} x_j \varepsilon_j + x_k \varepsilon_k + \sum_{k+1}^n x_j \varepsilon_j,$$

he finds the characteristic function for z and v, from which he derives $p(z, v)$. Using that $u = b - \beta = z/[xx]$ he proves that the joint distribution of (u, v) is asymptotically normal with zero mean and

$$\sigma_u^2 = \frac{\sigma^2}{[xx]}, \quad \sigma_v^2 = \frac{1}{4[xx]f^2(0)}, \quad \sigma_{uv} = \frac{\mu_{(1)}}{2[xx]f(0)}.$$

He writes the quadratic form in the exponent of $p(u, v)$ in the two ways that exhibit the two marginal and conditional distributions. Making the transformation $t = (1 - c)u + cv$, he proves that the minimum of $V(t)$ is obtained for

$$c = \frac{2f(0)\sigma^2(2f(0)\sigma^2 - \mu_{(1)})}{\sigma^2 - \mu_{(1)}^2 + (2f(0)\sigma^2 - \mu_{(1)})^2},$$

and that

$$\min_t V(t) = V(u)\frac{\sigma^2 - \mu_{(1)}^2}{\sigma^2 - \mu_{(1)}^2 + (2f(0)\sigma^2 - \mu_{(1)})^2}.$$

Laplace concludes that if $f(\varepsilon)$ is known and if $2f(0)\sigma^2 \neq \mu_{(1)}$ then we can find c and the estimate $b - c(b - \tilde{\beta})$ will be better than b. He points out that $2f(0)\sigma^2 = \mu_{(1)}$ for the normal distribution so in this case b cannot be improved.

Stigler [247] has compared Laplace's paper with Fisher's 1920 paper in which he introduces the concept of sufficiency. Fisher assumes that the observations are normally distributed and proves that s_1 and s_2 are asymptotically normal, that the efficiency of s_1 relative to s_2 is $1/(\pi - 2)$, and that s_2 is the best estimate of σ^2 among the moment estimates. He did not know that Gauss [102] had proved these results. Neither did he know that Laplace [162] had observed that a full comparison of the properties of two competing estimates has to be based on their joint distribution. However, as pointed out by Stigler, Fisher took two essential steps more than Laplace. He investigated $p(s_1, s_2)$ for a finite value of n, in the case $n = 4$, and showed that $p(s_1 | s_2)$ is independent of σ^2 and that the same property holds also for $s_3, s_4, \ldots$. Fisher concludes that "The whole of the information respecting σ^2, which a sample provides, is summed up in the value of $\sigma_2^2[s_2]$. This unique superiority of σ_2^2 is dependent on the form of the normal curve ....." The term sufficiency was introduced in his 1922 paper.

12.4 Generalizations of the Central Limit Theorem

There are two lines of research on the central limit theorem. The one develops conditions, as weak as possible, for the theorem to hold, and the other extends the theorem by considering the normal distribution as the main term of a series expansion.

As an example of the first method we state a theorem due to J.W. Lindeberg [166].

Let $x_1, x_2, \ldots$ be independent with distribution functions $F_1(x)$, $F_2(x)$, $\ldots$, $E(x_i) = 0$ and $V(x_i) = \sigma_i^2$. Setting $\gamma_n^2 = \sigma_1^2 + \cdots + \sigma_n^2$ and assuming that

$$\gamma_n^{-2} \sum_1^n \int_{|x| < t\gamma_n} x^2 dF_i(x) \to 1, \text{ for each } t > 0,$$

the normalized sum $u_n = (x_1 + \cdots + x_n)/\gamma_n$ is asymptotically normal $(0, 1)$.

Series expansions based on the normal distribution are used by Laplace [157] in his solution of a diffusion problem; see Hald ([114], § 17.8). Here we consider the expansion developed by Poisson [221] and Bienaymé [13].

To simplify the notation we introduce the cumulants instead of the moments. From (12.5) it follows that the characteristic function $\psi(t)$ is the moment generating function. Thiele [259] proposes to use $\ln \psi(t)$ as the cumulant generating function so that

$$\ln \psi(t) = \sum_{r=1}^{\infty} (it)^r \kappa_r / r!. \tag{12.15}$$

Comparing with (12.5) it can be seen that

$$\sum_{1}^{\infty} (it)^r \kappa_r/r! = \ln[1 + \sum_{1}^{\infty} (it)^r \mu'_r/r!].$$

(12.16)

Introducing the moments about the mean $\mu_r = E[(x - \mu'_1)^r]$ we get

$$\kappa_1 = \mu'_1, \ \kappa_2 = \mu_2, \ \kappa_3 = \mu_3, \ \kappa_4 = \mu_4 - 3\mu_2^2, \ \kappa_5 = \mu_5 - 10\mu_3\mu_2,$$
$$\kappa_6 = \mu_6 - 15\mu_4\mu_2 - 10\mu_3^2 + 30\mu_2^3.$$

Moreover, we need the Hermite polynomials defined by the relations

$$D_x^r \phi(x) = (-1)^r \phi(x) H_r(x), \ r = 0, 1, 2, \ldots.$$

(12.17)

Hence,

$$H_0(x) = 1, H_1(x) = x, H_2(x) = x^2 - 1, H_3(x) = x^3 - 3x,$$
$$H_4(x) = x^4 - 6x^2 + 3, H_5(x) = x^5 - 10x^3 + 15x,$$
$$H_6(x) = x^6 - 15x^4 + 45x^2 - 15.$$

The Hermite polynomials satisfy the orthogonality relation

$$\int_{-\infty}^{\infty} H_r(x) H_s(x) \phi(x) dx = \begin{cases} 0, & s \neq r \\ r!, & s = r. \end{cases}$$

With this modern notation we can give a compact derivation of the Poisson–Bienaymé expansion of $p(s_n), s_n = x_1 + \cdots + x_n$, for continuous variables by means of the inversion formula

$$p(s_n) = \frac{1}{2\pi} \int_{-\infty}^{\infty} \exp(-is_n t) \psi^n(t) dt.$$

The main term is

$$\frac{1}{2\pi} \int_{-\infty}^{\infty} \exp[i(n\kappa_1 - s_n)t - n\kappa_2 t^2/2] dt = (n\kappa_2)^{-1/2} \phi(u),$$

(12.18)

$u = (s_n - n\kappa_1)/\sqrt{n\kappa_2}$. To evaluate the following terms Poisson and Bienaymé differentiate (12.18) with respect to s_n with the result that

$$\frac{1}{2\pi} \int_{-\infty}^{\infty} \exp[i(n\kappa_1 - s_n)t - n\kappa_2 t^2/2](-it)^r dt$$
$$= (-1)^r (n\kappa_2)^{-(r+1)/2} H_r(u)\phi(u).$$

The following terms of the expansion are then easily found. Setting

$$\gamma_r = \kappa_{r+2}/\kappa_r^{(r+2)/2}, \ r = 1, 2, \ldots,$$

(12.19)

the expansion becomes

$$p(s_n) = \frac{\phi(u)}{\sqrt{n\kappa_2}} \left[1 + \frac{\gamma_1 H_3(u)}{3! n^{\frac{1}{2}}} + \frac{\gamma_2 H_4(u)}{4! n} + \frac{\gamma_3 H_5(u)}{5! n^{3/2}} \right.$$

$$\left. + \frac{1}{6!} \left(\frac{\gamma_4}{n^2} + 10 \frac{\gamma_1^2}{n} \right) H_6(u) + \cdots \right]$$

(12.20)

This expansion is today called the Gram–Charlier series. Ordering terms according to powers of $n^{-1/2}$, it is called the Edgeworth [46] series.

If the random variable is discrete with equidistant arguments, a continuity correction has to be introduced.

There exists a large literature on the central limit theorem and its extension. Surveys covering the period before 1900 are given by Adams [1] and Hald ([114], Chapter 17). Later results are discussed by Le Cam [23].

About 50 years later it was realized that frequency functions could be represented by orthogonal expansions analogous to the expansion of the sampling distribution above. The history of such expansions is given by Hald [115], [117].

13

Gauss's Theory of Linear Minimum Variance Estimation

13.1 The General Theory, 1823

Gauss's paper on the "Theory of the Combination of Observations Leading to Minimum Errors" was published in three parts, the first two in 1823 and a supplement in 1828.

Gauss points out that Laplace since 1811 had considered estimation from a new point of view by seeking the most advantageous combination of observations instead of the most probable value of the parameter, and that Laplace had proved that for large samples the best estimate is the least squares estimate regardless of the distribution of the errors. He states that the problem of finding the combination having the smallest error of estimation "is unquestionably one of the most important problems in the application of mathematics to the natural sciences." He proposes to supplement Laplace's theory by finding the linear combination that leads to the smallest mean square error of estimation for any sample size.

Whereas Laplace begins by discussing the properties of linear estimates in general and then proves that the best is the least squares estimate, Gauss reverses this procedure. First he introduces the normal equations and discusses the properties of the solution and afterward he compares with other linear estimates.

Using the same notation as in Section 7.3, Gauss's model is written as

$$y = X\beta + \varepsilon, \ E(\varepsilon) = 0, \ D(\varepsilon) = E(\varepsilon\varepsilon') = \sigma^2 I_n.$$

To solve the equation

$$X'X\beta = X'y + z, \ z = -X'\varepsilon,$$

Gauss sets

$$\beta = b + Qz,$$

where the unknowns, b and Q, are found by elimination of z, which leads to the equation

$$\beta = b + QX'X\beta - QX'y.$$

From this identity in β it follows that $QX'X = I_m$ and $b = QX'y$. That this solution agrees with the previous one is seen by multiplying the normal equations $X'Xb = X'y$ by Q.

To find the variance of the estimates, Gauss expresses b in terms of ε. Setting

$$b = A'y, \ A = XQ,$$

it follows that

$$A'X = I_m, \ A'A = Q,$$

and

$$b - \beta = -Qz = A'\varepsilon,$$

so

$$D(b) = E(A'\varepsilon\varepsilon'A) = \sigma^2 Q.$$

This is the first part of Gauss's proof of the properties of the least squares estimate: b is unbiased for β with dispersion matrix $\sigma^2 Q$, where Q is the inverse of $X'X$.

It remains to prove the optimality of b. As does Laplace, Gauss considers the class of linear error-consistent estimates

$$\tilde{\beta} = K'y = K'X\beta + K'\varepsilon, \ K = k_{is}, \ i = 1, \ldots, n, \ s = 1, \ldots, m,$$

for which $K'X = I_m$, so $\tilde{\beta} = \beta + K'\varepsilon$ and $D(\tilde{\beta}) = \sigma^2 K'K$. The problem is to minimize the diagonal elements of $K'K$. Noting that

$$\tilde{\beta} - b = (K - A)'\varepsilon = (K - A)'y - (K - A)'X\beta,$$

it follows that $(K - A)'X = 0$ because the left side does not depend on β. Hence

$$(K - A)'XQ = (K - A)'A = 0.$$

Setting $K = A + (K - A)$ we have

$$K'K = A'A + (K - A)'(K - A),$$

which shows that the diagonal elements of $K'K$ are minimized for $K = A$. This is the first general proof of the optimality of the least squares estimate. Laplace had proved the optimality for $m = 2$ only and in a more complicated manner.

Gauss remarks that the m equations $X'XQ' = I_m$ may be solved by the same method as the normal equations $X'Xb = X'y$ because the matrix of coefficients is the same. The estimate and its dispersion matrix may thus be found by one compact numerical procedure.

If the observations have different variances, $D(\varepsilon) = \sigma^2 P^{-1}$, where P is a known diagonal matrix with positive diagonal elements, then the least squares estimates are found by minimizing $(y - X\beta)'P(y - X\beta)$ and $D(b) = \sigma^2(X'PX)^{-1}$.

Assuming that the observations are normally distributed and using inverse probability, Gauss [100] had proved that $E(\beta) = b$ and $D(\beta) = \sigma^2 Q$, so that the credibility limits for β_r are $b_r \pm u\sigma\sqrt{q_{rr}}$. In 1823 he proved that $E(b) = \beta$ and $D(b) = \sigma^2 Q$ under the weak assumption that $E(\varepsilon) = 0$ and $D(\varepsilon) = \sigma^2 I_n$. It is peculiar that he did not mention that the confidence limits for β_r under the assumption of normality are the same as the credibility limits.

In 1809 he had not discussed the estimation of σ^2. In 1823 he proved that $s^2 = [ee]/(n - m)$ is an unbiased estimate of σ^2 and that

$$V(s^2) = \frac{\mu_4 - \sigma^2}{n - m} - \frac{\mu_4 - 3\sigma^4}{(n - m)^2} \left(m - \sum_{i=1}^{n} h_{ii}^2 \right), \quad H = XQX',$$

tacitly assuming that μ_4 is finite.

It is very remarkable, said Gauss, that under normality $V\{[\varepsilon\varepsilon]/n\} = 2\sigma^4/n$ and $V\{[ee]/(n - m)\} = 2\sigma^4/(n - m)$, which shows that $[ee]$ can be regarded as a sum of $n - m$ squared independent errors.

After these fundamental results Gauss discussed some special problems of great practical importance. We mention some results without giving the proofs. Let h denote a vector of real numbers of dimension m. The best estimate of $h'\beta$ is $h'b$ and $V(h'b) = \sigma^2 h'Qh$.

Under the linear restriction $h'(b - \beta) = \gamma$, where γ is a given constant, the best estimate of β is

$$\hat{\beta} = b - \frac{\gamma Qh}{h'Qh},$$

and

$$\min_{\beta} \varepsilon'\varepsilon = e'e + \frac{\gamma^2}{h'Qh}, \quad e = y - Xb.$$

Gauss solves the problem of updating the least squares estimates by an additional observation $y_0 = h'\beta + \varepsilon_0$, say. Introducing the "residual" $e_0 = y_0 - h'b$, the least squares estimate based on the $n + 1$ observations equals

$$\hat{b} = b + \frac{e_0 Qh}{1 + h'Qh},$$

$$D(\hat{b}) = \sigma^2 \left(Q - \frac{Qhh'Q}{1 + h'Qh} \right),$$

and

$$\min_{\beta}(\varepsilon'\varepsilon + \varepsilon_0^2) = e'e + \frac{e_0^2}{1 + h'Qh}.$$

He gives an analogous formula for the effect of changing the weight of one of the observations.

13.2 Estimation Under Linear Constraints, 1828

In the Supplement Gauss discusses another version of the linear model inspired by his work in geodesy, where the parameters are subject to linear constraints. In linearized form the problem is to estimate β in the model $y = \beta + \varepsilon$ under the r linearly independent restrictions $F'\beta = f$, where F is an $(n \times r)$ matrix of known constants and f a known vector of dimension $r < n$. Gauss remarks that this model may be transformed to the previous one by using the restrictions to eliminate r of the n parameters so there remains $m = n - r$ freely varying parameters. However, if $r < n/2$ it is simpler to use the new model directly.

To find the best estimate of $\theta = h'\beta$ Gauss writes the class of linear estimates as

$$\tilde{\theta} = h'y - \alpha'(F'y - f) = (h - F\alpha)'y + \alpha'f,$$

where α is an arbitrary vector to be determined such that

$$V(\tilde{\theta}) = \sigma^2 (h - F\alpha)'(h - F\alpha)$$

is minimized. The solution, a say, is therefore obtained by solving the normal equations $F'Fa = F'h$, which leads to the estimate

$$t = h'y - a'(F'y - f),$$

and

$$V(t) = \sigma^2(h'h - h'Fa),$$

because $F'(h - Fa) = 0$.

Writing t in the form $h'b$ it follows that

$$y = b + e, \ \ e = FR(F'y - f), \ \ RF'F = I_r.$$

Because $e = FRF'\varepsilon$ and

$$b - \beta = \varepsilon - e = (I_r - FRF')\varepsilon,$$

Gauss finds that

$$\varepsilon'\varepsilon = e'e + (b - \beta)'(b - \beta),$$

so $\varepsilon'\varepsilon$ is minimized for $\beta = b$.

He proves that $E(e'e) = r\sigma^2$ and thus $E\{(b - \beta)'(b - \beta)\} = (n - r)\sigma^2$, and states without proof the variance of $e'e/r$.

Hence, Gauss solves all the problems of estimation for the linear model of full rank. His results and the ideas in his proofs can be found today in many textbooks on estimation, the first model under the name of regression analysis and the second as analysis of variance.

13.3 A Review of Justifications for the Method of Least Squares

By 1823 statistical justifications for using the method of least squares had been derived from the following four methods of estimation:

(1) Gauss [100] combined a normal error distribution with a uniform prior distribution and defined the best estimate as the value maximizing the posterior density.

(2) Using that the arithmetic mean in large samples from an error distribution with finite variance is normally distributed and assuming that the prior distribution is uniform, Laplace [156] defined the best estimate as the value minimizing the expected absolute error in the posterior distribution.

(3) Using that linear functions of observations from an error distribution with finite variance are multivariate normal in large samples and considering the class of linear unbiased estimates, Laplace [157] defined the best estimate as the one having minimum expected absolute error or equivalently minimum variance.

(4) Assuming that the error distribution has finite variance and considering the class of linear unbiased estimates, Gauss [103] defined the best estimate as the one having minimum variance.

The first two proofs use inverse probability and lead to posterior probability intervals for the parameters. The last two proofs use direct probability, that is, the sampling distribution of the estimates. Because of the asymptotic normality of the estimates, Laplace could find large-sample confidence intervals for the parameters. For small samples Gauss could not do so because the error distribution was unspecified. However, it is implicit in Gauss's paper that the 3σ-limits will give a large confidence coefficient.

After 1812 Laplace and Gauss preferred the frequentist theory. This is obvious from Laplace's Supplements to the TAP and his later papers and from Gauss's papers ([103],[104]) and his letter (1839) to Bessel, although Gauss never publicly said so and continued to use the first proof in his introductory lectures on the method of least squares; see his letter (1844) to Schumacher. Nevertheless, in most textbooks occurring between 1830 and 1890, Gauss's first proof is used as motivation for the method of least squares, presumably because his first proof is much simpler than the second.

In his writings on estimation theory, Laplace often expressed the opinion that a priori we are ordinarily ignorant of the mathematical form of the error distribution, an opinion accepted by Gauss. They therefore made the weakest possible assumption on the error distribution, namely the existence of the second moment. Despite this general attitude Laplace admitted that under special circumstances it is reasonable to assume a normal error distribution, and he noted that his asymptotic results then hold for any sample size.

It is strange that Laplace and Gauss, who estimate laws of nature by means of statistical methods, did not study empirical distributions to find out the common forms for laws of error. However, this was done by Bessel [10], who found that the errors of astronomical observations under typical conditions are nearly normally distributed. One may wonder why Gauss did not react to this new information by supplementing his second proof by an exposition of the sampling theory for the linear normal model. In this way he could have presented a comprehensive theory covering linear unbiased minimum variance estimation under both weak and strong conditions.

The advantage of assuming a normal error distribution is of course that exact confidence limits for the parameters may be obtained if σ^2 is known. However, the next step would be to find the confidence coefficient for the limits

$$b_r \pm ts\sqrt{q_{rr}},$$

which Laplace had derived for large samples in the First Supplement [161] to the TAP.

Gauss did not attempt to solve this problem, which required a derivation of the t-distribution, but he made two contributions to its solution. First, he introduced the number of degrees of freedom for the residual sum of squares, replacing Laplace's large-sample estimate $(e'e)/n$ by the unbiased estimate $(e'e)/(n-m)$. Next, he remarked that under normality the sum of the n squared residuals may be considered as a sum of $n-m$ independent squared errors. However, he did not go on to say that this implies that the two terms of the decomposition

$$\varepsilon'\varepsilon = e'e + (b-\beta)'X'X(b-\beta)$$

are independent and that the second term is distributed as a sum of m squared errors.

The logical continuation of these considerations is to observe the distribution of the variance ratio

$$F = \frac{[(b-\beta)'X'X(b-\beta)]/m}{(e'e)/(n-m)}.$$

This problem was formulated and solved by Fisher [71] about a century after Gauss's contribution.

13.4 The State of Estimation Theory About 1830

For a student of mathematical statistics in 1830, the theory of estimation must have been a confusing topic because of the many conflicting methods that had been proposed. Only a few (if any) managed to read and understand all of Laplace and Gauss. Inasmuch as our previous exposition is encumbered with many technical details, we here attempt to give an overview of the main ideas as a basis for the following discussion.

Let us first summarize Laplace's asymptotic theory which consists of two theorems of great generality, based on inverse and direct probability, respectively.

First, the posterior distribution of the parameter θ is asymptotically normal $(\hat{\theta}, \sigma_\theta^2)$, where $\hat{\theta}$ is the mode and

$$\sigma_\theta^{-2} = -D_\theta^2 \ln p(\hat{\theta}|\underline{x}). \tag{13.1}$$

For two parameters, (θ_1, θ_2) is asymptotically normal with mean $(\hat{\theta}_1, \hat{\theta}_2)$ and inverse dispersion matrix

$$\{\sigma^{ij}\} = \left\{ -\frac{\partial^2 \ln p(\hat{\theta}_1, \hat{\theta}_2 | \underline{x})}{\partial \theta_i \partial \theta_j} \right\}, \ i, j = 1, 2.$$

The proof follows simply from the Taylor expansion of $\ln p(\theta_1, \theta_2 | \underline{x})$ around the mode and is easily generalized to the multiparameter case, as remarked by Laplace. This theorem is the basis for his solutions of estimation and testing problems for binomially and multinomially distributed variables.

For estimating the location parameter in a symmetric error distribution, he did not appeal to the result above but proved afresh, by developing $\ln p(\theta|\underline{x})$ in a Taylor series around the mode, that θ is asymptotically normal with mean $\hat{\theta}$ and variance (13.1). Because the mode and the median coincide for a symmetric distribution, he remarks that $\hat{\theta}$ equals the posterior median and as such minimizes the expected absolute error.

He does not explain why he prefers the mode for binomially distributed observations but the median in error theory, unless one takes his appeal to expected loss as an argument applicable only to errors.

Laplace's second asymptotic result is the multivariate central limit theorem. He proved that the two linear forms $[w_1 \varepsilon]$ and $[w_2 \varepsilon]$ of the n errors $\varepsilon_1, \ldots, \varepsilon_n$ are asymptotically normal with covariance matrix $\{\sigma^2[w_i w_j]\}, i, j = 1, 2, \sigma^2$ denoting the error variance, irrespective of the form of the error distribution. He used this result to prove that the linear minimum variance estimate of the parameters in the linear model is obtained by the method of least squares and states the generalization to the multiparameter case.

For finite samples, Laplace attacked the problem of estimating the location parameter from both points of view but without success. For the traditional estimate, the arithmetic mean, he derived the sampling distribution by means of the convolution formula. Using inverse probability, he proposed the posterior median as estimate. Both procedures, applied to the then known error distributions, led to complicated calculations so they could be used only for very small samples.

The breakthrough came with Gauss's invention of the normal distribution. Combining Laplace's principle of inverse probability with the posterior mode as the estimate, Gauss found the arithmetic mean as the estimate of the location parameter and showed that the posterior distribution of θ is normal $(\overline{x}, \sigma^2/n)$ for any sample size. Replacing θ with a linear combination of m parameters, he gave the first probabilistic proof of the method of least squares. Because of its intuitive appeal and mathematical simplicity, this proof came to enjoy wide popularity.

Between 1811 and 1828 Laplace and Gauss developed their frequentist theory of linear estimation for the linear model $y = X\beta + \varepsilon$, assuming that $E(\varepsilon) = 0, D(\varepsilon) = \sigma^2 I_n$, and $0 < \sigma^2 < \infty$. Laplace required the estimate of β to be linear, error-consistent (today replaced by unbiased), and of minimum variance in large samples. He showed that the least squares estimate satisfies these requirements. Gauss proved that this estimate has minimum variance for any sample size.

Laplace used $[ee]/n$ as estimate of σ^2 in large samples and derived its variance. Gauss improved this result showing that $[ee]/(n-m)$ is unbiased for σ^2 and found its variance for any sample size.

It was essential for both Laplace and Gauss that the minimum variance property holds regardless of the form of the error distribution. If the error distribution is symmetric, a lower bound for the confidence coefficient can be obtained from Gauss's inequality.

For the special case of a normal error distribution, the linear estimate is normally distributed, and exact confidence limits can be found for any sample size if the variance is known, as shown by Laplace in the Second and Third Supplements to the TAP. However, his proof is based on minimization of the variance of the estimate; he does not maximize the probability density of the sample or, as we would say today, the likelihood function. As pointed out by Laplace, the restriction to linear estimates is essential from a computational point of view.

The price to be paid for the linearity and the robustness of the method of least squares is a loss of efficiency in the nonnormal case. This was investigated by Laplace by comparing the method of situation with the method of least squares. In the simple case $y_i = \beta + \varepsilon_i$, he found that the sample median minimizes $\sum |y_i - \beta|$, whereas the sample mean minimizes $\sum (y_i - \beta)^2$. He derived the distribution of the sample median and showed that asymptotically the median has a smaller variance than the mean for error distributions more peaked than the normal. Hence a nonlinear estimate may be more efficient than a linear estimate.

It is clear that Laplace and Gauss after 1812 preferred the frequentist theory of linear estimation to the previous inverse probability theory. The requirement of minimum variance was a natural consequence of the fact that linear estimates asymptotically are normally distributed.

Without explicitly discussing the contradictions involved in the existing estimation theory, Laplace realized the need for a common principle. Having previously rejected the posterior mode, he now also rejected the median and proposed instead to use the posterior mean, presumably because it minimizes the expected squared error of the posterior distribution, just as he in the frequentist theory had minimized the expected squared error of the sampling distribution. Because the median and the mean coincide, for a normal distribution this change did not affect the estimate of the location parameter, so we have to look at the estimate of the squared precision for a demonstration.

Assuming that the squared precision of a normal distribution, $k = 1/2\sigma^2$, is uniformly distributed, and using the posterior mean as estimate, Laplace found [161] $[\varepsilon\varepsilon]/(n+2)$ and later ([159] 3rd ed.) $[ee]/(n+1)$ as estimates of σ^2. In both cases his comment is that "the value of k which should be chosen is evidently the integral of the products of the values of k multiplied by their probabilities."

He criticizes Gauss [102] for using the posterior mode to derive the estimate $[\varepsilon\varepsilon]/n$ under the assumption that $h = \sqrt{k}$ is uniformly distributed.

The contradictions were never openly discussed. However, Gauss [105] in a letter to Bessel remarked that if he had been as wise in 1809 as he was in 1839 he would (like Laplace) have used the posterior mean instead of the mode, but he never said so publicly. In the same letter he distanced himself from the principle of inverse probability by characterizing it as "metaphysical."

It is therefore no wonder that the followers of Laplace and Gauss had difficulties in deciding whether to use frequentist or inverse probability theory, and in the latter case whether to use the mode or the mean.

Part IV

ERROR THEORY. SKEW DISTRIBUTIONS. CORRELATION. SAMPLING DISTRIBUTIONS

The Development of a Frequentist Error Theory

14.1 The Transition from Inverse to Frequentist Error Theory

Gauss did not take the trouble to rewrite his first proof of the method of least squares in terms of direct probability. This task was carried out by astronomers and geodesists writing elementary textbooks on the method of least squares. They found Gauss's second proof too cumbersome for their readers and did not need the generalization involved because the measurement errors encountered in their fields were in most cases nearly normally distributed. As far as error theory is concerned, they realized that the principle of inverse probability was superfluous. The method of maximizing the posterior density could be replaced by the method of maximizing the density $p(\underline{x}|\underline{\theta})$ of the observations, which would lead to the same estimates because $p(\underline{x}|\underline{\theta}) \propto p(\underline{\theta}|\underline{x})$. This method has an obvious intuitive appeal and goes back to Daniel Bernoulli and Lambert; see Todhunter ([261], pp. 236–237) and Edwards [50]. Todhunter writes:

> Thus Daniel Bernoulli agrees in some respects with modern theory. The chief difference is that modern theory takes for the curve of probability that defined by the equation
>
> $$y = \sqrt{c/\pi}\,e^{-cx^2},$$
>
> while Daniel Bernoulli takes a [semi]circle.

The astronomers considered only the error model, assuming that the errors

$$\varepsilon_i = x_i - g_i(\theta), \quad i = 1, \ldots, n,$$

are symmetrically distributed about zero. Replacing the true value by its linear Taylor approximation and assuming that errors are normally distributed with precision constant h they got the linear normal model. It is important to note that their terminology differs from today's in two respects. For the probability of an error to fall in the interval $(\varepsilon, \varepsilon + d\varepsilon)$ they write $f(\varepsilon)d\varepsilon$ so that the corresponding probability of the observed system of errors equals

$$f(\varepsilon_1) \cdots f(\varepsilon_n)d\varepsilon_1 \cdots d\varepsilon_n = Pd\varepsilon_1 \cdots d\varepsilon_n.$$

However, they called P the *probability* of the system of errors, the term "probability density" being of a later date. We have

$$P = f(x_1 - g_1(\underline{\theta}), h) \cdots f(x_n - g_n(\underline{\theta}), h).$$

Their method of estimating $\underline{\theta}$ consisted of maximizing P with respect to $\underline{\theta}$, and for a given value of $\underline{\theta}$ they estimated h by the same method. They called the resulting estimates "the most probable values of the unknowns." This is of course a misuse of the word probable because their model implied that the parameters are unknown constants, and not random variables as in inverse probability. This terminological confusion was not cleared up until Fisher [66] introduced the term likelihood for $p(\underline{x}|\underline{\theta})$ as a function of $\underline{\theta}$ for a given value of $\underline{x}$. In their notation they did not distinguish between the true value and its estimate.

For the linear normal model, the method of maximizing P with respect to $\underline{\theta}$ leads to the method of least squares. The sampling distribution and the optimality of the least squares estimates follow from the theory of linear minimum variance estimation by Laplace and Gauss.

J.F. Encke [56] wrote a comprehensive survey of Gauss's work on the method of least squares. He reproduced Gauss's derivation of the normal distribution based on inverse probability and the principle of the arithmetic mean. He (p. 276) continued:

> ... the joint probability of the coincidence of n errors in these observations is $p(\underline{x}|\theta, h)$. This probability becomes largest, when the sum of the squares of the remaining errors after an assumed hypothesis [regarding θ] is the smallest possible, and consequently will *the hypothesis about θ leading to the absolute minimum of the remaining errors* be the most probable among all possible hypotheses also according to Theorem II [the principle of inverse probability].

Encke thus begins by maximizing the probability of the observations, calling the estimate the most probable, and afterwards he notes that the same estimate is obtained by maximizing the posterior probability.

14.2 Hagen's Hypothesis of Elementary Errors and His Maximum Likelihood Argument, 1837

In his textbook for civil engineers G.H.L. Hagen [112] begins by deriving the normal distribution of errors by a simplification of Laplace's central limit theorem. He thus avoids using the axiom of the arithmetic mean and inverse probability as in Gauss's first proof. He formulates the hypothesis of elementary errors as follows ([112], p. 34).

> The hypothesis, which I make, says: the error in the result of a measurement is the algebraic sum of an infinitely large number of elementary errors which are all equally large, and each of which can be positive or negative with equal ease.

He notes that this formulation is a simplification of the real measurement process, obtained by replacing the positive errors in a symmetric distribution by their mean, and similarly for the negative errors. This means that the distribution of the sum of n elementary errors is the symmetric binomial, which converges to the normal for $n \to \infty$. To avoid the complicated proofs of de Moivre and Laplace he finds the relative slope of the binomial frequency curve, which for $n \to \infty$ leads to a differential equation with the normal distribution as solution. Because of its simplicity this proof became popular.

Assuming that the errors of the observations are normally distributed and setting each of the n differentials of the errors equal to $d\varepsilon$, he (p. 67) gets (in our notation)

$$p(\underline{\varepsilon})d\underline{\varepsilon} = (d\varepsilon/\sqrt{\pi})^n \exp(-h^2[\varepsilon\varepsilon]).$$

He remarks that

> The first factor of this expression will be unchanged even if we attach another hypothesis [regarding the true value] to the observations and the individual errors consequently take on other values; the second factor will however be changed. Among all hypotheses of this kind, which can be attached to the observations, the most probable is consequently the one which makes $Y[p(\underline{\varepsilon})d\underline{\varepsilon}]$ a maximum, which means that the exponent of e should be a minimum, that is, the sum of the squares of the resulting errors should be as small as possible.

Hagen's second factor is thus the likelihood function which he maximizes to find the most likely hypothesis.

For the linear model $\varepsilon = y - X\beta$ he gets

$$[\varepsilon\varepsilon] = [ee] + (b - \beta)'X'X(b - \beta),$$

which inserted into $p(\underline{\varepsilon})$ gives the likelihood function for β. To find the likelihood for β_1, he (p. 80) maximizes $p(\underline{\varepsilon})$ with respect to the other βs and finds

$$\max_{\beta_2,\dots,\beta_m} p(\underline{\varepsilon}) \propto \exp(-h^2(b_1 - \beta_1)^2/q_{11}), \ Q = (X'X)^{-1}.$$

He concludes that $V(b_r) = \sigma^2 q_{rr}$, $r = 1,\dots,m$, and gives the explicit expression for q_{rr} in terms of the elements of $X'X$ for $m = 1, 2, 3$.

If he had used inverse or direct probability, he should have found the marginal distribution by integration.

Hagen writes as if he has found the variance in the sampling distribution of b_r, but with hindsight we can see that it is the curvature of the likelihood function. For the linear model we thus have three methods leading to the same "probability limits" for β_r: (1) the posterior distribution due to Gauss [100]; (2) the sampling distribution due to Laplace ([157], [161]); and Gauss [103], and (3) the likelihood function due to Hagen [112].

14.3 Frequentist Theory, Chauvenet 1863, and Merriman 1884

In his textbook on astronomy W. Chauvenet [27] wrote an appendix on the method of least squares, which essentially is an abridged English version of Encke's 1832 paper but leaving out all material on inverse probability. He reproduces Encke's proof of the arithmetic mean as the most probable estimate of the true value, and interpreting "most probable" as the maximum of $p(\underline{x}|\theta)$ with respect to θ he uses the same mathematics as Gauss to derive the normal distribution without mentioning $p(\theta|\underline{x})$. After having stated the probability density P for the sample, he (pp. 481–482) remarks that

> The most probable system of values of the unknown quantities [...] will be that which makes the probability P a maximum.

Specializing to the normal distribution the method of least squares follows.

Maximizing $p(\underline{x}|\theta, h)$ with respect to h, he finds $n/2[\varepsilon\varepsilon]$ as estimate of h^2 and using that $[\varepsilon\varepsilon] = [ee] + n(\overline{x} - \theta)^2$ he gets $(n-1)/2[ee]$. He thus succeeds in proving Gauss's basic results for normally distributed observations by operating on the likelihood function instead of the posterior distribution.

A more consistent exposition of this theory is due to Merriman [176] in *The Method of Least Squares*, written for "civil engineers who have not had the benefit of extended mathematical training." Merriman had an extraordinarily good background for writing this book because he in 1877 had provided a valuable "List of Writings Relating to the Method of Least Squares with Historical and Critical Notes," containing his comments on 408 books and papers published between 1722 and 1876. It should be noted, however, that all his comments are based on the principle of maximizing the probability of the sample; he does not even mention inverse probability.

Merriman begins with the classical definition of probability but changes to the frequency definition in Art. 17:

> The probability of an assigned accidental error in a set of measure-ments is the ratio of the number of errors of that magnitude to the total number of errors.

In Art. 13 he defines "most probable" as follows. "The most probable event among several is that which has the greatest mathematical probability."

He gives two derivations of the normal distribution. First, he reports Hagen's demonstration based on the hypothesis of elementary errors, and next he simplifies Gauss's proof, pointing out that he uses the word "most probable" for the arithmetic mean in the sense of Art. 13.

For two unknowns, he (Art. 28) gives the joint probability density P of the observations and states that "the most probable values of the unknown quantities are those which render P a maximum (Art. 13)." In Art. 41 he writes similarly:

> The most probable system of errors will be that for which P is a maximum (Art. 13) and the most probable values of the unknowns will correspond to the most probable system of errors.

This postulate obviously leads to the maximum likelihood estimate disguised as the most probable value of the unknown. Applying this principle to normally distributed errors, the method of least squares follows.

To estimate h Merriman (Art. 65) says that the probability of the occurrence of n independent errors equals

$$P' = \pi^{-n/2} h^n \exp(-h^2 [\varepsilon\varepsilon]) (d\varepsilon)^n .$$

> Now, for a given system of errors, the most probable value of h is that which has the greatest probability; or h must have such a value as to render P' a maximum.

This leads to the estimate $\sqrt{n/2 [\varepsilon\varepsilon]}$, which he modified to $\sqrt{(n-1)/2 [ee]}$.

Merriman (Art. 164) finds the uncertainty of the estimate $\hat{h} = \sqrt{n/2 [\varepsilon\varepsilon]}$ from the formula

$$p(\underline{x}|\theta, \hat{h}(1+\delta)) = p(\underline{x}|\theta, \hat{h}) e^{-n\delta^2} (1 + O(\delta)).$$

He concludes that the standard deviation of δ is $1/\sqrt{2n}$, so that the standard error of $\hat{h}$ equals $\hat{h}/\sqrt{2n}$. This is the likelihood version of Gauss's 1816 proof.

15

Skew Distributions and the Method of Moments

15.1 The Need for Skew Distributions

During the period from about 1830 to 1900 statistical methods gradually came to be used in fields other than the natural sciences. Three pioneers were Quetelet (anthropometry, social sciences), Fechner (psychophysics, factorial experiments), and Galton (genetics, biology, regression, correlation). Applications also occurred in demography, insurance, economics, and medicine. The normal distribution, originally introduced for describing the variation of errors of measurement, was used by Quetelet and Galton to describe the variation of characteristics of individuals. However, in many of the new applications skew distributions were encountered that led to the invention of systems of nonnormal distributions.

It was natural to start by "modifying" the normal distribution. Thiele and Gram used the first two terms of the Gram–Charlier series as a skew distribution, and Thiele introduced $\kappa_3/\kappa_2^{3/2}$ as a measure of skewness and κ_4/κ_2^2 as a measure of kurtosis, κ denoting the cumulants. Fechner combined two normal distributions with common mode and different standard deviations. Galton noted that if height is normally distributed, weight cannot be so and proposed to consider the logarithm of such measures as normal. Independently, Thiele, Edgeworth, and Kapteyn generalized this idea by taking a suitably selected function of the observations as normally distributed.

Hagen had derived the normal distribution by solving a differential equation analogous to the difference equation satisfied by the binomial. Generalizing this approach Pearson derived a four-parameter system of distributions by solving a differential equation of the same form as the difference equation satisfied by the hypergeometric distribution. He used the same measures of skewness and kurtosis as Thiele but expressed in terms of moments.

To estimate the parameters in the new distributions, several new methods were developed, which were simpler than the method of least squares. Galton used two percentiles to fit a normal distribution to his data, and Kapteyn similarly used four percentiles to estimate the parameters in his model. Thiele used empirical cumulants as estimates of the theoretical cumulants and Pearson used empirical moments as estimates of the theoretical moments. Expressing the theoretical quantities as functions of the parameters and solving for the parameters they found the estimates.

For a detailed analysis of the works of Quetelet, Fechner, Lexis, and Galton, we refer to Stigler [250].

15.2 Series Expansions of Frequency Functions. The A and B Series

The expansion (12.20) of $p(s_n)$ took on a new significance when Hagen [112] and Bessel [11] formulated the hypothesis of elementary errors, saying that an observation may be considered as a sum of a large number of independent elementary errors stemming from different sources and with different unknown distributions. Hence, s_n is interpreted as an observation and $p(s_n)$ as the corresponding frequency function. A difficulty with this interpretation is the fact that we do not know the measuring process (or other processes considered) in such detail that we can specify the number of elementary errors making up an observation, so it is only the form of $p(s_n)$ that is known. Hagen and Bessel therefore used the expansion only as an argument for considering the normal distribution as a good approximation to empirical error distributions.

However, Thiele and Gram went further and considered expansions of frequency functions of the form

$$g(x) = \sum_{j=0}^{\infty} c_j f_j(x), \ c_0 = 1, \ -\infty < x < \infty, \tag{15.1}$$

where $g(x)$ is a given frequency function and $f(x) = f_0(x)$ another frequency function chosen as a first approximation to $g(x)$. It is assumed that $g(x)$ and its derivatives or differences tend to zero for $|x| \to \infty$. In the discussion of such series there are several problems involved: (1) the choice of $f_0(x)$; (2) the relation of $f_j(x)$, $j \geq 1$, to $f_0(x)$; and (3) the determination of c_j.

In the following it is assumed that the moment generating functions of $g(x)$ and $f_j(x)$ exist. Denoting the moments of $g(x)$ by μ_r and the "moments" of $f_j(x)$ by ν_{rj} it follows that the cs may be expressed in terms of the moments by solving the linear equations

$$\mu_r = \sum_{j=0}^{\infty} c_j \nu_{rj}, \ r = 1, 2, \ldots. \tag{15.2}$$

This formula is valid for both continuous and discontinuous distributions. The solution is commonly simplified by choosing the fs such that the matrix $\{\nu_{rj}\}$ is lower triangular, which means that c_j becomes a linear combination of $\mu_1, \ldots, \mu_j$.

Another approach consists of choosing the fs as orthogonal with respect to the weight function $1/f_0(x)$ and using the method of least squares, which gives

$$c_j = \int [f_j(x)/f_0(x)] \, g(x)dx \; / \int \left[f_j^2(x)/f_0(x) \right] dx. \qquad (15.3)$$

If $f_j(x) = f_0(x)P_j(x)$, where $P_j(x)$ is a polynomial of degree j, then c_j becomes proportional to $E[P_j(x)]$, which is a linear combination of the first j moments of $g(x)$. Hence, this special case leads to the same result as the special case of (15.2).

For an appropriate choice of the fs, the first few terms of the series will often give a good approximation to $g(x)$. However, the partial sum

$$g_m(x) = \sum_{j=0}^{m} c_j f_j(x), \; m = 1, 2, \ldots, \qquad (15.4)$$

will not necessarily be a frequency function; $g_m(x)$ may, for example, take on negative values.

If $g(x)$ is continuous, Thiele and Gram use the normal density as $f_0(x)$ and its jth derivate as $f_j(x)$. The resulting series is called the (normal) A series.

If $g(x)$ is discontinuous, Lipps uses the Poisson frequency function as $f_0(x)$ and its jth difference as $f_j(x)$. This series is called the (Poisson) B series.

The terms A and B series were introduced by Charlier [24], who studied the expansion (15.1) for arbitrary continuous and discontinuous frequency functions. When Charlier wrote his first papers on this topic he was not aware of the fact that the normal A series and the Poisson B series previously had been discussed by several authors.

Several other authors derived the two series by other methods; the history has been written by Hald (2002). The present exposition is limited to the works of the three pioneers, Thiele, Gram, and Lipps, with some comments on the contributions of Charlier and Steffensen.

T.N. Thiele (1838–1910) got his master's degree in mathematics and astronomy at the University of Copenhagen in 1860 and his doctor's degree in astronomy in 1866. After having worked for ten years as assistant at the Copenhagen Observatory, he was in 1870–1871 employed in establishing the actuarial basis for the life insurance company Hafnia, that was started in 1872 with Thiele as actuary. In 1875 he became Professor of Astronomy and Director of the Copenhagen Observatory. He kept up his relation to Hafnia, but from 1875 with J.P. Gram as collaborator. Thiele worked in astronomy, numerical analysis, actuarial mathematics, and applied and mathematical statistics. Most of his contributions to statistics are contained in the three textbooks *Almindelig Iagttagelseslære* (*The General Theory of Observations*) 1889, *Elementær Iagttagelseslære* (*The Elementary Theory of Observations*) 1897, and a slightly revised English version *Theory of Observations* 1903, reprinted in the *Annals of Mathematical Statistics*, 1931. A translation with commentaries of the 1889 book is given by Lauritzen [164].

J.P. Gram (1850–1916), mathematician, actuary, and statistician, gives in his doctoral thesis [110] a comprehensive account (in Danish) of the series expansion of an "arbitrary" function in terms of various systems of orthogonal functions with discussion of the convergence problem and with statistical applications, in particular to the expansion of skew probability densities. An abridged German version, mainly on the convergence problem was published in 1883. Gram's basic idea is that the least squares estimates in the linear model should not be affected by adding new independent variables. He invented a new method for solving the normal equations by introducing linear orthogonal combinations of the original vectors.

C.V.L. Charlier (1862–1934) studied at the University of Uppsala, where he in 1887 got his doctor's degree in astronomy and became associate professor. In 1897 he became professor of astronomy at the University of Lund. His analysis of astronomical observations led him in 1905 to mathematical statistics on which he published a large number of papers. He had his own series of publications from the Astronomical Observatory so neither the originality nor the quality of his papers were checked by referees. His textbook *Grunddragen af den matematiska statistiken,* 1910, was published in a German version as *Vorlesungen über die Grundzüge der mathematischen Statistik,* 1920.

G.F. Lipps (1865–1931) studied mathematics, physics, and philosophy at the universities of Leipzig and München. He got his doctor's degree in mathematics in 1888 and in philosophy in 1904. He became professor of philosophy and psychology at the University of Zürich in 1911. After Fechner's death his incomplete manuscript to the important work *Kollektivmasslehre* was edited and completed by Lipps in 1897. His main work in statistics is the 215-page-long paper *Die Theorie der Collektivgegenstände* [167] that was published as a book the following year. It is a good textbook on mathematical statistics which contains a rather complete discussion of the *A* and *B* series.

J.F. Steffensen (1873–1961) studied law at the University of Copenhagen and graduated in 1896. After several administrative jobs he became a member of the newly established State Insurance Board in 1904. At the same time he studied astronomy, mathematics, and statistics on his own and got his doctor's degree in mathematics in 1912. In 1919 he became associate professor of actuarial mathematics and in 1923 full professor. He wrote three excellent textbooks, *Matematisk Iagttagelseslære* (*The Mathematical Theory of Observations*) 1923, *Interpolationslære* 1925 with an English edition in 1927 and *Forsikringsmatematik* (*Actuarial Mathematics*) 1934.

The fundamental theory of the *A* series is due to Gram ([110], [111]) who writes the series in the form

$$g(x) = f(x) \sum_{j=0}^{\infty} c_j P_j(x), \qquad (15.5)$$

where $\{P_j(x)\}$ are orthogonal polynomials satisfying the relation

$$\int P_j(x)P_k(x)f(x)dx = 0 \text{ for } j \neq k.$$

He determines c_j by the method of least squares using $1/f(x)$ as weight function, which leads to

$$c_j = \int P_j(x)g(x)dx \Big/ \int P_j^2(x)f(x)dx, \ j = 0, 1, \ldots . \quad (15.6)$$

Because $P_j(x)$ is a polynomial of degree j it follows that c_j is a linear combination of the moments of $g(x)$ of order 1 to j. As special cases he discusses the A series with the normal and the gamma distributions as leading terms.

Thiele [257] gives a simple derivation of the normal A series, which he writes as

$$g(x) = \sum_{j=0}^{\infty}(-1)^j \frac{1}{j!}c_j D_x^j \phi(x) = \phi(x)\sum_{j=0}^{\infty}\frac{1}{j!}H_j(x). \quad (15.7)$$

Multiplying by $H_k(x)$, integrating, and using the orthogonality of the Hs he finds $c_j = E[H_j(x)]$, which gives c_j in terms of the moments of $g(x)$. Introducing the cumulants instead of the moments and replacing x by the standardized variable $u = (x - \kappa_1)\kappa_2^{-1/2}$ he gets the series in the final form as

$$g(x) = \kappa_2^{-1/2}\phi(u)[1 + \gamma_1 H_3(u)/3! + \gamma_2 H_4(u)/4! + \gamma_3 H_5(u)/5!$$

$$+(\gamma_4 + 10\gamma_1^2)H_6(u)/6! + \cdots], \quad (15.8)$$

$$\gamma_r = \kappa_{r+2}/\kappa_2^{(r+2)/2}, \ r = 1, 2, \ldots .$$

This series has the same form as (12.20). In fact the extended central limit theorem may be obtained by replacing x by $\bar{x}$.

Turning to the discontinuous case we discuss Lipp's [167] derivation of the Poisson B series. Assuming that $g(x) = 0$ for $x < 0$ he writes the B series as

$$g(x) = \sum_{j=0}^{\infty}c_j \nabla^j f(x), \ \nabla f(x) = f(x) - f(x-1), \ x = 0, 1, \ldots . \quad (15.9)$$

Setting

$$f(x) = e^{-\lambda}\lambda^x/x!, \ \lambda > 0,$$

he finds

$$\nabla^j f(x) = f(x)P_j(x), \ j = 0, 1, \ldots ,$$

where

$$P_j(x) = \sum_{k=0}^{j}(-1)^k \binom{j}{k}\lambda^{-k}x^{(k)}, \quad (15.10)$$

$x^{(k)} = x(x-1)\cdots(x-k+1)$, $k \geq 1$, and $x^{(0)} = 1$.

Instead of the ordinary moments he introduces the binomial moments, which we denote by α and β, respectively. Multiplying (15.9) by $\binom{x}{r}$ and summing over x, he gets

$$\alpha_r = \sum_{j=0}^{r} \beta_{rj} c_j, \quad r = 0, 1, \ldots, \tag{15.11}$$

where

$$\alpha_r = \sum_{x} \binom{x}{r} g(x)$$

and

$$\beta_{rj} = \sum_{x} \binom{x}{r} \nabla^j f(x) = (-1)^j \lambda^{r-j}(r-j)!, \quad j = 0, 1, \ldots, r. \tag{15.12}$$

Hence, the matrix of coefficients in (15.11) is lower triangular, and solving for c_r, Lipps gets

$$c_r = \sum_{j=0}^{r} (-1)^j [(r-j)!]^{-1} \lambda^{r-j} \alpha_j. \tag{15.13}$$

Looking at Lipp's proof from the Thiele–Gram point of view, it can be seen that Lipp's series may be written as

$$g(x) = f(x) \sum_{j=0}^{\infty} c_j P_j(x).$$

C. Jordan [131] proves that the Ps are orthogonal,

$$\sum_{x} P_j(x) P_k(x) f(x) = 0 \text{ for } j \neq k,$$

and

$$\sum_{x} P_j^2(x) f(x) = r! \lambda^{-r}.$$

Using orthogonality, it follows that

$$c_r = \frac{1}{r!} \lambda^r \sum_{x} P_r(x) g(x),$$

which by means of (15.10) immediately gives Lipp's result (15.13).

Steffensen looks at the problem of series expansion of frequency functions from a purely statistical point of view. He [245] writes:

We are therefore of opinion that the labour that has been expended by several authors in examining the conditions under which the A-series is ultimately convergent, interesting as it is from the point of view of mathematical analysis, has no bearing on the question of the statistical applications.

The statistical problem is, he says, to fit a frequency function, $g_m(x)$ say, containing m parameters to a sample of n observations, $m < n$, and therefore the series has to be finite. Moreover, $g_m(x)$ should be a probability distribution, that is, it should be nonnegative and its sum or integral over the whole domain should be unity.

As an expert in the calculus of finite differences he derives a formula for $g_m(x)$ as a linear combination of the differences of $f(x)$ up to order m, the differences being defined as

$$\nabla_\alpha f(x) = [f(x) - f(x - \alpha)]/\alpha.$$

The B series then follows for $\alpha = 1$ and the A series for $\alpha \to 0$.

Steffensen concludes:

> There are, however, considerable drawbacks. We cannot, as with Pearson's types, be sure beforehand that negative values will not occur; as a matter of fact they often do occur, and this can only be ascertained at the end of the calculation. We have not even very good reason to expect that by adding another term such negative value will be made to disappear. (...) We are therefore inclined to think that the apparent generality of (28) [his general formula, containing the A and B series] is rather a disadvantage than otherwise, and that Pearson's types are as a rule preferable.

Considering the normal A series based on the first four moments, Barton and Dennis [2] have determined the region in which, as functions of the moments, the series is nonnegative.

15.3 Biography of Karl Pearson

The life of Karl Pearson (1857–1936) may be divided into three periods: before 1892, 1892–1922, and after 1922. He was born in London as the son of a barrister and studied mathematics in Cambridge 1875–1879. He got a fellowship at King's College, which made him financially independent for a number of years and which he used for traveling and studying. In Germany he studied physics, metaphysics, Darwinism, Roman law, German folklore, and German history, in particular the history of the Reformation. He also studied law in London and was called to the bar in 1881, but he practiced only for a short time. Pearson rebelled against the norms of Victorian society by being a freethinker, a socialist, a Darwinist, and a supporter of eugenics and feminism.

Besides lecturing and writing on these matters, he produced a large number of books and papers on pure and applied mathematics and physical science. In 1884 he became professor of applied mathematics and mechanics at University College London, and from 1891 to 1894 he also lectured at Gresham College, which resulted in *The Grammar of Science* [199] in which he expounded his views on the fundamental concepts of science. This marks the end of the first period of his scientific life.

Under the influence of Galton and Weldon, a drastic change in Pearson's scientific interests took place about 1892. But how could an applied mathematician, who knew next to nothing about statistical methods, help Weldon with the analysis of zoological data with the purpose of elucidating Darwin's theory of evolution? At the age of 35 Pearson began to educate himself in mathematical statistics, at the same time analyzing Weldon's data and developing new methods with an astounding energy and speed. He read a number of continental textbooks on social and demographic statistics, among them H. Westergaard's *Die Grundzüge der Theorie der Statistik* [266], which he considered the best on the relation between statistics and probability. Westergaard had parameterized the normal approximation to the binomial by means of $\sqrt{npq}$; Pearson ([200], p. 80) did the same and introduced the term "standard deviation."

In his first statistical papers Pearson derived the moments of the binomial and dissected an asymmetrical frequency curve into two normal curves. However, the two models are too limited in scope and, like Chebyshev, Fechner, Thiele, and Gram, Pearson felt the need for a collection of continuous distributions for describing the biological phenomena he was studying. He wanted a system embracing distributions with finite as well as infinite support and with skewness both to the right and the left. The breakthrough came with his famous paper, "Skew Variation in Homogeneous Material" [201] in which he derived the four-parameter system of continuous densities, that we discuss in the next section.

Another breakthrough is his [203] derivation of the χ^2 test for goodness of fit, which we discuss in Section 15.5.

He wrote a useful survey "On the Probable Errors of Frequency Constants," [204], which indicates his conversion from inverse probability; see Pearson and Filon [213], on the frequentist theory.

His lifelong studies on the theory of correlation and regression with applications to heredity, eugenics, and biology begin with "Regression, Heredity, and Panmixia" [202], another important memoir in this series is "On the General Theory of Skew Correlation and Nonlinear Regression" [206]. In parallel he developed a theory of multiple contingency, beginning with "On the Theory of Contingency and Its Relation to Association and Normal Correlation" [205]. It is characteristic for all his papers that the theory is illustrated by a wealth of applications to anthropological, biological, and demographic data. However, in many cases the examples illustrate the application of the new statistical techniques rather than contributing to a deeper understanding of the subject matter in question.

To further the applications of his methods, he published auxiliary tables of statistical functions in *Biometrika*. Important collections of such tables are *Tables for Statisticians and Biometricians*, Part I [207] and Part II [210], with introductions on their use. Furthermore, he edited *Tables of the Incomplete Γ-Function* [209], *Tables of the Incomplete Beta-Function* [211], and a series of *Tracts for Computers*, beginning in 1919.

His burden of work was enormous. He founded and edited *Biometrika, A Journal for the Statistical Study of Biological Problems* (1901) together with Weldon and in consultation with Galton, he lectured on mathematical statistics and he founded a Biometric Laboratory. In 1906 he also took over Galton's Eugenics Laboratory. Finally, in 1911 he was relieved of his duties as professor of applied mathematics by becoming the first Galton professor of eugenics and head of a department of applied statistics to which his Biometric and Eugenics Laboratories were transferred.

Between 1892 and 1911 he thus created his own kingdom of mathematical statistics and biometry in which he reigned supremely, defending its ever-expanding frontiers against attacks. He retired in 1934, but about 1922 he was succeeded by Fisher as the leading British statistician.

Among Pearson's many contributions after 1922 are the completion of a biography of Galton and *The History of Statistics in the 17th and 18th Centuries Against the Changing Background of Intellectual, Scientific, and Religious Thought*, edited by E.S. Pearson [212]. It is a valuable companion to Todhunter's *History*. In this period he also founded and edited the *Annals of Eugenics* from 1925.

An annotated *Bibliography of the Statistical and Other Writings of Karl Pearson* [186] is due to G.M. Morant with the assistance of B.L. Welsh. It contains 648 items; 406 are on the theory of statistics and its applications.

Pearson was not a great mathematician but he effectively solved problems head-on by elementary methods. His proofs are detailed and lengthy.

He was a fighter who vigorously reacted against opinions that seemed to detract from his own theories. Instead of giving room for other methods and seeking cooperation his aggressive style led to controversy.

The most comprehensive biography, *Karl Pearson: An Appreciation of Some Aspects of his Life and Work* is due to his son E.S. Pearson [191], who later ([192], [193], [194]) wrote on important aspects of the history of biometry and statistics in the period in question. A survey mainly of Pearson's statistical work is due to Eisenhart [53]; MacKenzie [172] discusses Pearson's life and work in a social context.

Pearson was unduly criticized by his following generation; see Fisher ([86], pp. 2–4). A more balanced view is expressed at the end of E.S. Pearson's biography:

...but having himself provided a mathematical technique and a system of auxiliary tables, by ceaseless illustration in all manner of problems he at least convinced his contemporaries that the employment of this novel calculus [of mathematical statistics] was a practical proposition. From this has resulted a permanent change which will last, whatever formulae, whatever details of method, whatever new conceptions of probability may be employed by coming generations in the future.

15.4 Pearson's Four-Parameter System of Continuous Distributions, 1895

Pearson defines the moments as

$$\mu'_r = E(x^r), \ r = 0, 1, \ldots, \ \mu_r = E[(x - \mu'_1)^r], \ r = 2, 3, \ldots$$

and

$$\beta_1 = \mu_3^2/\mu_2^3, \ \beta_2 = \mu_4/\mu_2^2, \ \beta_3 = \mu_3\mu_5/\mu_2^4.$$

He derives the normal distribution from the symmetric binomial

$$p(x) = \binom{n}{x} \left(\frac{1}{2}\right)^n,$$

by calculating the relative slope of the frequency curve

$$S = \frac{p(x+1) - p(x)}{\frac{1}{2}\left[p(x+1) + p(x)\right]} = -\frac{x + \frac{1}{2} - \frac{1}{2}n}{(n+1)/4},$$

from which he concludes that the corresponding continuous distribution satisfies the differential equation

$$\frac{d\ln p(x)}{dx} = -\frac{x - \frac{1}{2}n}{n/4}.$$

The solution is the normal distribution with mean $n/2$ and variance $n/4$. A similar proof is due to Hagen [112] and Westergaard [266].

Pearson then analyzes the skew binomial and the hypergeometric distribution in the same way. For the latter he finds

$$p(x) = \binom{n}{x} \frac{(Np)^{(x)} (Nq)^{n-x}}{N^{(n)}}, \tag{15.14}$$

which gives

$$S = -y/(\beta_1 + \beta_2 y + \beta_3 y^2), \ y = \left(x + \frac{1}{2} - \mu\right),$$

where μ, β_1, β_2, and β_3 are constants depending on the parameters of (15.14). Hence, the corresponding continuous density satisfies a differential equation of the form

$$\frac{d\ln p(x)}{dx} = -\frac{x-\alpha}{\beta_1 + \beta_2 x + \beta_3 x^2}. \tag{15.15}$$

The solution depends on the sign of $\beta_2^2 - 4\beta_1\beta_3$, and Pearson discusses in great detail the resulting distributions which he classifies into several types.

We give a summary of his results based on Elderton [54], which is the standard text on this topic.

Writing p for $p(x)$, Pearson's system is based on the differential equation

$$\frac{d\ln p(x)}{dx} = \frac{x+a}{b_0 + b_1 x + b_2 x^2}. \tag{15.16}$$

It follows that

$$x^r(b_0 + b_1 x + b_2 x^2)p' = x^r(x+a)p,$$

which after integration gives

$$-b_0\mu'_{r-1} - (r-1)b_1\mu'_r - (r+2)b_2\mu'_{r+1} = \mu'_{r+1} + a\mu'_r, \quad r = 0, 1, \dots \tag{15.17}$$

under the assumption that $x^r(b_0 + b_1 x + b_2 x^2)p$ vanishes at the endpoints of the support for p. Hence, there is a one-to-one correspondence between a, b_0, b_1, b_2 and the first four moments, so p is uniquely determined from the first four moments.

The solution depends on the roots of the equation

$$b_0 + b_1 x + b_2 x^2 = 0, \tag{15.18}$$

that is, on $b_1^2/4b_0 b_2$, which expressed in terms of the moments gives the criterion

$$\kappa = \frac{\beta_1(\beta_2 + 3)^2}{4(2\beta_2 - 3\beta_1 - 6)(4\beta_2 - 3\beta_1)}. \tag{15.19}$$

Pearson distinguishes among three main types depending on whether $\kappa < 0$, $0 < \kappa < 1$, or $\kappa > 1$. In the first case the roots of (15.18) are real and of different sign, in the second the roots are complex, and in the third they are real and of the same sign. The corresponding distributions are Pearson's Types I, IV, and VI. Besides the main types he derives a number of transition types for $\kappa = 0$ and $\kappa = 1$, among them the normal and the gamma distribution. The value of κ, [i.e., of (β_1, β_2)], thus determines the type. A survey of the resulting system is shown in Table 15.1.

The empirical moments based on a sample of n observations are defined as

$$m'_r = \frac{1}{n}\sum_{i=1}^n x_i^r, \quad r = 0, 1, \dots, \quad \text{and} \quad m_r = \frac{1}{n}\sum_{i=1}^n (x_i - m'_1)^r, \quad r = 2, 3, \dots.$$

Table 15.1. Table of Pearson's Type I to VII distributions *Source*: E.S. Pearson and H.O. Hartley ([195], p. 79), slightly modified.

Type	Equation $y =$	Origin for x	Limits for x	Criterion
I	$y_0\left(1+\frac{x}{a_1}\right)^m\left(1-\frac{x}{a_2}\right)^m$	Mode	$-a_1 \leq x \leq a_2$	$\kappa < 0$
II	$y_0\left(1-\frac{x^2}{a^2}\right)^m$	Mean (=mode)	$-a \leq x \leq a,$	$\kappa = 0$
III	$y_0 e^{-\gamma x}(1+\frac{x}{a})^{\gamma a}$	Mode	$-a \leq x < \infty,$	$\kappa = \infty$
IV	$y_0 e^{-v\tan^{-1} x/a}(1+\frac{x^2}{a^2})^{-m}$	Mean $+\frac{va}{r}$, $r = 2m-2$	$-\infty < x < -\infty, 0 < \kappa < 1$	
V	$y_0 e^{-\gamma/x}x^{-p}$	At start of curve	$0 \leq x < \infty,$	$\kappa = 1$
VI	$y_0(x-a)^{q_2}x^{-q_1}$	At or before start of curve	$a \leq x < \infty,$	$\kappa > 1$
VII	$y_0\left(1+\frac{x^2}{a^2}\right)^{-m}$	Mean(=mode)	$-\infty < x < \infty,$	$\kappa = 0$

They are unbiased estimates of the theoretical moments. Pearson estimates the parameters in his distributions by the method of moments, which consists of setting $m'_r = \mu'_r$ for $r = 1,2,3,4$, and solving for the parameters.

Pearson ended the paper by fitting his distributions to a variety of data ranging from meteorology, anthropometry, zoology, botany, economics, demography, and to mortality statistics, a total of 15 examples. Hence, he not only provided a new collection of distributions, he also demonstrated their ability to fit actual data. It turned out, however, that the paper also raised many problems that were solved gradually during the next 30 years or so. We mention the most important.

(1) How can the goodness of fit be measured objectively? Pearson [203] solves this problem by deriving the χ^2 goodness-of-fit test, discussed in the next section.

(2) How does grouping affect the moments? Let $\lambda_2, \lambda_3, \ldots$ be moments about the mean for a grouped distribution with a grouping interval of length h. By means of a relation between integrals and sums, Sheppard [234] proves that μ_r with good approximation may be obtained from λ_r by the following corrections: $\mu_2 = \lambda_2 - h^2/12$, $\mu_3 = \lambda_3 - \lambda_1 h^2/4$, $\mu_4 = \lambda_4 - \lambda_2 h^2/2 + 7h^4/240$, where $\lambda_1 = \mu'_1$; see Hald [116] for the history of this topic.

(3) Is the method of moments an efficient method of estimation? Fisher [67] shows that this is not so in general, and as an example he proves that for symmetric Pearson distributions the method of moments has an efficiency of 80 percent. or more if β_2 lies between 2.65 and 3.42, whereas outside this

interval the efficiency is lower. For the normal distribution $\beta_2 = 3$ and the efficiency is 100 percent.

(4) Steffensen ([243], [245]) points out that there is no probabilistic interpretation of Pearson's differential equation because the restrictions on the parameters of the hypergeometric are not carried over to Pearson's parameters. He gives a probabilistic interpretation of Pearson's Type I based on a roulette experiment and shows that the other types, apart from Type IV, can be derived from Type I.

(5) What is the purpose of fitting a Pearson distribution to empirical data? Pearson does not discuss this problem; he is content with the fact that a good graduation is obtained. However, the advantage of characterizing the data by a few estimated parameters becomes obvious when several sets of data of the same kind are to be compared. The standard error of the moments required for this purpose was provided by Sheppard [235].

15.5 Pearson's χ^2 Test for Goodness of Fit, 1900

In the first part of Pearson's paper [203] on the χ^2 test, he assumes that the k-dimensional random variable z is multivariate normal with mean zero and dispersion matrix D so that

$$p(z) \propto \exp(-\frac{1}{2}z'D^{-1}z).$$

He remarks that

$$\chi^2 = z'D^{-1}z$$

represents a generalized ellipsoid in the sample space and that $P(\chi > \chi_0)$ may be found by integrating $p(z)$ over the corresponding region. He notes that the ellipsoid by a linear transformation may be turned into a sphere so that

$$P(\chi > \chi_0) = \frac{\int \cdots \int\limits_{\chi_0 < \chi < \infty} e^{-\chi^2/2}dt_1 \cdots dt_k}{\int \cdots \int\limits_{0 < \chi < \infty} e^{-\chi^2/2}dt_1 \cdots dt_k},$$

$t_1, \ldots, t_k$ being the new coordinates. He continues:

> Now suppose a transformation of coordinates to generalized polar coordinates, in which χ may be treated as the ray, then the numerator and the denominator will have common integral factors really representing the generalized "solid angles" and having identical limits.

He thus obtains

$$P(\chi > \chi_0) = \frac{\int_{\chi_0}^{\infty} e^{-\chi^2/2}\chi^{k-1}d\chi}{\int_0^{\infty} e^{-\chi^2/2}\chi^{k-1}d\chi}. \tag{15.20}$$

The details of the proof may be found in Kendall and Stuart ([135], § 11.2).

Pearson concludes that if z and D are known, then we can calculate χ^2 and

> ...an evaluation of (15.20) gives us what appears to be a fairly reasonable criterion of the probability of such an error [or a larger one] occurring on a random selection being made.

He begins the second part of the paper by stating his objective as follows. "Now let us apply the above results to the problem of fit of an observed to a theoretical frequency distribution." He considers a multinomial distribution with class probabilities $p_1, \ldots, p_k$, $\sum p_i = 1$, a sample of size n with $x_1, \ldots, x_k$ observations in the k classes, $\sum x_i = n$, and the deviations $e_i = x_i - np_i$, $i = 1, \ldots, k$, $\sum e_i = 0$. He proves that the limiting distribution of the statistic

$$\chi^2 = \sum_{i=1}^{k} \frac{e_i^2}{np_i}$$

for $n \to \infty$ is the χ^2 distribution with $k - 1$ degrees of freedom.

In the following we use the term "degrees of freedom" for the number of cells minus the number of independent linear restrictions on the frequencies, although this term was not introduced in the present context until Fisher [68]. We denote the number of degrees of freedom by f.

Without proof Pearson states that the variance of e_i equals np_iq_i and the covariance of e_i and e_j equals $-np_ip_j$. A proof is given by Sheppard [235]. Moreover he assumes that $(e_1 \ldots, e_{k-1})$ is asymptotically normal: $e_k = -(e_1 + \cdots + e_{k-1})$. He then makes a trigonometrical transformation that is equivalent to introducing the new variables

$$y_i = \frac{e_i}{p_i\sqrt{n}}, \quad i = 1, \ldots, k, \quad \sum p_i y_i = 0.$$

Hence the dispersion matrix of $(y_1, \ldots, y_{k-1})$ is

$$A = \begin{bmatrix} \frac{q_1}{p_1} & -1 & \cdots & -1 \\ -1 & \frac{q_2}{p_2} & \cdots & -1 \\ \vdots & \vdots & & \vdots \\ -1 & -1 & \cdots & \frac{q_{k-1}}{p_{k-1}} \end{bmatrix}.$$

Because of the simple structure of A, it is easy to evaluate A^{-1}. Pearson finds that

$$a^{ii} = p_i + p_i^2 p_k^{-1} \text{ and } a^{ij} = p_i p_j p_k^{-1}, \quad i \neq j,$$

so

$$\sum_{1}^{k-1}\sum_{1}^{k-1} a^{ij} y_i y_j = \sum_{1}^{k-1}\sum_{1}^{k-1} \frac{a^{ij} e_i e_j}{n p_i p_j}$$

$$= \sum_{1}^{k-1} \frac{e_i^2}{n p_i} + \sum_{1}^{k-1}\sum_{1}^{k-1} \frac{e_i e_j}{n p_k} = \sum_{1}^{k} \frac{e_i^2}{n p_i}.$$

Pearson points out that in practice the ps are seldomly known; we usually estimate the parameters in the theoretical frequency function from the sample so instead of $p_i = p_i(\theta)$ we have to use $\hat{p}_i = p_i(\hat{\theta})$, where $\hat{\theta}$ denotes the estimated parameters. He wrongly concludes that the effect of this substitution is negligible. This led to a long dispute with Fisher [68] who introduced the number of degrees of freedom for the χ^2 test, defined as the number of cells in the multinomial minus the number of independent linear restrictions on the frequencies.

15.6 The Asymptotic Distribution of the Moments by Sheppard, 1899

A complete discussion of the asymptotic distribution of empirical moments is due to Sheppard [235]. Let $z = f(x_1, \ldots, x_k)$ be a differentiable function of k random variables with finite moments. Using Taylor's expansion, Sheppard proves Gauss's formula $V(z) \cong f_0' D f_0$, where f_0 denotes the vector of derivatives $\partial f / \partial x_i$, $i = 1, \ldots, k$, taken at the true value of the xs, and D denotes the dispersion matrix of the xs. This method for finding the variance of a differentiable function is called the δ-method. A similar formula holds for the covariance of two functions of the same variables.

Sheppard assumes that the observations come from a discrete distribution or a grouped continuous distribution. Let the corresponding multinomial have class probabilities $p_1, \ldots, p_k$, $\sum p_i = 1$, and let a sample of size n have $n h_1, \ldots, n h_k$ observations in the k classes; $\sum h_i = 1$. Sheppard proves that $E(h_i) = p_i$, $n V(h_i) = p_i q_i$, and $n CV(h_i, h_j) = -p_i p_j$, $i \neq j$. It follows that the linear form $z = \sum \alpha_i (h_i - p_i)$ is asymptotically normal with zero mean and that

$$nV(z) = \sum \alpha_i^2 p_i - \left(\sum \alpha_i p_i\right)^2. \tag{15.21}$$

Because this holds for any linear function, he concludes that $(h_1, \ldots, h_k)$ are normally correlated.

Let $x_1, \ldots, x_k$ denote deviations from the true mean and set $p(x_i) = p_i$ so that $E(x) = \sum x_i p_i = 0$,

$$\mu_t = \sum x_i^t p_i, \text{ and } m_t = \sum (x_i - \bar{x})^t h_i, \ t = 1, 2, \ldots.$$

Using the binomial theorem Sheppard gets

$$m_t = \sum (x_i^t - t\overline{x}x_i^{t-1} + \cdots)h_i = \mu_t + \sum (x_i^t - t\mu_{t-1}x_i)(h_i - p_i) + \cdots.$$

Hence, $m_t - \mu_t$ is approximately a linear function of the deviations $h_i - p_i$ and using (15.21) Sheppard finds

$$nV(m_t) = \mu_{2t} - 2t\mu_{t-1}\mu_{t+1} + t^2\mu_{t-1}^2\mu_2 - \mu_t^2.$$

By the same method he derives the variances and covariances of the moments and gives similar results for the bivariate moments

$$m_{st} = \frac{1}{n}\sum (x_i - \overline{x})^s(y_i - \overline{y})^t, \ (s,t) = 0,1,2,\ldots.$$

By means of the δ-method he derives the large-sample variance of the correlation coefficient $r = m_{11}(m_{20}m_{02})^{-1/2}$, which for the bivariate normal becomes $V(r) = (1 - \rho^2)^2/n$. Pearson [204] used Sheppard's results and the δ-method to find the large-sample variances and covariances of the estimates of the parameters in his system of distributions.

Sheppard also found the variances of Galton's percentile estimates of the parameters in the normal distribution; he determined the optimum choice of percentiles and discussed the efficiency of these estimates in relation to the moment estimates. He generalized this analysis by studying the properties of linear combinations of percentiles and the corresponding estimates.

15.7 Kapteyn's Derivation of Skew Distributions, 1903

J.C. Kapteyn (1851–1922), professor of astronomy at Groningen, The Netherlands, writes in the preface to his book [132] that

> I was requested by several botanical students and by some other persons interested in the statistical methods of Quetelet, Galton, Pearson ..., to deliver a few lectures in which these methods would be explained in a popular way. In studying the literature on the subject, in order to meet this request to the best of my ability, I soon found that, not only would it be extremely difficult, if not impossible, to present Pearson's theory of skew curves to non-mathematical hearers in such a form that they might be enabled to apply it in their work, but that the theory itself was open to grave objections. I was thus led to an independent investigation of the subject.

About Pearson's system he further remarks that

> ...it does not connect the form of the observed curve with the causes to which this form is due, so that no insight whatever can be gained in the nature of these causes.

Kapteyn then develops a theory of the genesis of frequency curves inspired by growth processes of plants and animals. By reversing this reasoning he proposes to characterize the growth process from the form of the frequency curve.

An extended version of his theory was published in 1916 with M.J. van Uven 1878–1959), professor of mathematics at Wageningen, as co-author. There Kapteyn writes that "the conclusions to which the theory leads must not be taken as well established facts but rather as 'working hypotheses'."

He maintains that all distributions in nature are skew, and that the reason for the successful applications of the normal distribution is the fact that in these cases the standard deviation is so small compared with the mean that the skewness becomes negligible.

He estimates the parameters by equating the empirical and theoretical percentiles.

We give a simplified version of his proof.

Let an element of magnitude (quality) ξ_0 be subjected to a process that successively alters the expected magnitude of ξ_0 to $\xi_1, \xi_2, \ldots$, corresponding to the different phases of the process. The change in magnitude at the ith phase, $\xi_i - \xi_{i-1}$, is assumed to depend on a "cause" acting with intensity η_i, and the magnitude of the element ξ_{i-1}, in the following manner

$$\xi_i - \xi_{i-1} = \eta_i h(\xi_{i-1}); \tag{15.22}$$

that is, the change in the magnitude of the element is proportional to the product of the intensity of the cause and a function of the magnitude of the element when the cause starts to act. η_i is called the reaction intensity, and $h(\xi)$ the reaction function.

The changes in the magnitude of the element during the first n phases may be characterized by the equations

$$\begin{aligned}
\xi_1 &= \xi_0 + \eta_1 h(\xi_0), \\
\xi_2 &= \xi_1 + \eta_2 h(\xi_1), \\
&\ \vdots \\
\xi_n &= \xi_{n-1} + \eta_n h(\xi_{n-1}).
\end{aligned} \tag{15.23}$$

In order to determine $\xi_n - \xi_0$ as a function of $\eta_1, \eta_2, \ldots, \eta_n$, (15.22) is written

$$\eta_i = \frac{\xi_i - \xi_{i-1}}{h(\xi_{i-1})},$$

and hence

$$\sum_{i=1}^{n} \eta_i = \sum_{i=1}^{n} \frac{\xi_i - \xi_{i-1}}{h(\xi_{i-1})}.$$

Assuming that the number of causes influencing the final result is large and the changes in magnitude at every stage comparatively small, we have

$$\sum_{i=1}^{n} \eta_i = \sum_{i=1}^{n} \frac{\xi_i - \xi_{i-1}}{h(\xi_{i-1})} \simeq \int_{\xi_0}^{\xi_n} \frac{dx}{h(x)}. \tag{15.24}$$

If we introduce

$$\xi_n = \sum_{i=1}^{n} \eta_i$$

and

$$g(\xi) = \int_{\xi_0}^{\xi} \frac{dx}{h(x)},$$

(15.24) may be written

$$\xi_n = g(\xi_n), \tag{15.25}$$

it now being possible to determine the size of the element at the end of the nth phase by solving (15.25) with respect to ξ_n.

In practical work it is not possible to keep the conditions of the process constant, and at each phase the reaction intensity will therefore deviate from the above stated theoretical values, and the changes in magnitude of the elements, partaking in the process, will vary. Assuming that the reaction intensity at the ith phase, y_i, is a random variable with mean value η_i and variance σ_i^2, the changes in magnitude of a given element may be characterized by the following equations, equivalent to (15.23)

$$x_1 = x_0 + y_1 h(x_0), \ x_0 = \xi_0,$$
$$x_2 = x_1 + y_2 h(x_1),$$
$$\vdots$$
$$x_n = x_{n-1} + y_n h(x_{n-1}).$$

In analogy with (15.24) we get

$$\sum_{i=1}^{n} y_i = \sum_{i=1}^{n} \frac{x_i - x_{i-1}}{h(x_{i-1})} \simeq \int_{x_0}^{x_n} \frac{dx}{h(x)}$$

and

$$\sum_{i=1}^{n} y_i = z_n = g(x_n). \tag{15.26}$$

According to the central limit theorem z_n will be normally distributed under certain general conditions when $n \to \infty$, the mean being ζ_n. Equation (15.26) then implies that the elements will not be normally distributed according to size, but that a function, $g(x)$, of the size will be normally distributed. For $h(x) = 1$, we find that $g(x) = x - x_0$; that is, x is normally distributed. Thus, if the reaction function is constant, which means that the changes in magnitude are independent of the size already obtained when the causes start to act, then the distribution according to size will be normal.

If the reaction function $h(x)$ is equal to x, we have

$$g(x) = \int_{x_0}^{x} \frac{dx}{x} = \ln x - \ln x_0;$$

that is, $\ln x$ is normally distributed. Thus, if the change in magnitude corresponding to a given cause is proportional to the intensity of that cause and further to the size of the element, the distribution obtained will be logarithmic normal.

16

Normal Correlation and Regression

16.1 Some Early Cases of Normal Correlation and Regression

We employ the modern notation of the multivariate normal distribution. Let $x = (x_1, \ldots, x_m)$ be a vector of normally correlated random variables with density

$$p(x) = (2\pi)^{-m/2} |A|^{1/2} \exp(-\frac{1}{2}(x - \mu)' A (x - \mu)), \qquad (16.1)$$

where μ is the vector of expectations and A a positive definite $m \times m$ matrix.

Defining the variances and covariances as

$$\sigma_{ij} = E(x_i - \mu_i)(x_j - \mu_j), \ \sigma_{ii} = \sigma_i^2, \ (i, j) = 1, \ldots, m,$$

and the dispersion matrix $D = \{\sigma_{ij}\}$, it may be proved that $D = A^{-1}$, so

$$p(x) = (2\pi)^{-m/2} |D|^{-1/2} \exp(-\frac{1}{2}(x - \mu)' D^{-1} (x - \mu)). \qquad (16.2)$$

Finally, introducing the standardized variables $u_i = (x_i - \mu_i)/\sigma_i$, the correlation coefficients

$$\rho_{ij} = E(u_i u_j) = \frac{\sigma_{ij}}{\sigma_i \sigma_j}, \ \rho_{ii} = 1,$$

and the corresponding matrix $C = \{\rho_{ij}\}$, we have

$$p(u) = (2\pi)^{-m/2} |C|^{-1/2} \exp(-\frac{1}{2} u' C^{-1} u). \qquad (16.3)$$

Laplace [157] obtains the bivariate normal as the asymptotic distribution of two linear forms of errors in the form (16.2). His investigation of the efficiency of two estimation methods [162] is based on the conditional distributions of the estimates in the bivariate normal.

A. Bravais (1811–1863), a French naval officer, who later became professor of astronomy and physics, writes [22] on "the probabilities of the errors of the position of a point." He considers m linear combinations of n normally distributed errors, $z_r = [k_r \varepsilon]$, $r = 1, \ldots, m$, $m \le n$, where ε_i is normal $(0, \sigma_i^2)$. To find the distribution of the zs, Bravais supplements $z_1, \ldots, z_m$ by $n - m$ linearly independent functions, $z_{m+1}, \ldots, z_n$, say, so that $z = K'\varepsilon$, where K is an $n \times n$ matrix. Solving for ε, he finds $\varepsilon = (K')^{-1}z$,

$$p(z) = g(\varepsilon) \left| \frac{\partial \varepsilon}{\partial z} \right|, \text{ where } g(\varepsilon) = \prod_1^n (\sqrt{2\pi}\sigma_i)^{-1} \exp(-\varepsilon_i^2/2\sigma_i^2),$$

and integrating out the auxiliary variables, he proves that z is multivariate normal for $m = 2$ and 3. He writes that a similar result probably holds for $m > 3$ but he has not been able to prove so.

For $m = 2$ he obtains $p(z_1, z_2)$ in the form (16.2), which he transforms to (16.1) as

$$p(z_1, z_2) = (c/\pi)e^{-q}, \ c^2 = a_{11}a_{22} - a_{12}^2, \ q = a_{11}z_1^2 + 2a_{12}z_1z_2 + a_{22}z_2^2,$$

where the as depends on the ks and the σs. He points out that the equation obtained by setting the exponent equal to a constant defines an ellipse and that the corresponding values of (z_1, z_2) thus have equal probability density. He transforms the quadratic form to a sum of squares by an orthogonal transformation. He derives analogous results for $m = 3$ and shows that $2q$ is distributed as χ^2 with two and three degrees of freedom, respectively.

In connection with his discussion of the properties of the contour ellipse, Bravais determines the horizontal tangent by differentiating the quadratic form for a given value of q. He finds that the value of z_1 corresponding to the maximum value of z_2 is given by the relation $z_1 = -a_{12}z_2/a_{11}$, which in modern notation may be written as $z_1 = (\rho\sigma_1/\sigma_2)z_2$, the regression of z_1 on z_2. He shows the regression line in a graph together with the ellipse and its horizontal tangent.

Bravais's paper gives the first systematic treatment of the mathematical properties of the two- and three-dimensional normal distribution. These properties are implicit in the writings of Laplace and Gauss, but they did not give a systematic account.

Bienaymé [13] generalizes Laplace's proof of the central limit theorem. For the linear normal model $y = X\beta + \varepsilon$, β may be estimated by $b = K'y$, where K is an $(n \times m)$ matrix satisfying $K'X = I_m$. It follows that b is normal $(\beta, K'K\sigma^2)$. Bienaymé's multivariate central limit theorem says that the above result for b holds for $n \to \infty$ regardless of the distribution of ε if only the variance is finite.

He criticizes Laplace and Gauss for using confidence intervals for single parameters only and proposes to use confidence ellipsoids instead. Under the normality assumption he considers the quadratic form

$$z'z = (b - \beta)'(K'K)^{-1}(b - \beta)/2\sigma^2,$$ (16.4)

and proves by iteration that

$$P(z'z < c^2) = \frac{2}{\Gamma(m/2)} \int_0^c t^{m-1} e^{-t^2} dt,$$ (16.5)

which shows that $2z'z$ is distributed as χ^2 with m degrees of freedom. For $P = \frac{1}{2}$ and $m = 2$ he compares the marginal confidence intervals with the confidence ellipse.

In his *Ausgleichungsrechnung*, Helmert ([119], pp. 231–256) treats the problem of "error" within the framework of Gauss's linear model and shows how the axes and the regression lines depend on the coefficient matrix of the normal equations. He illustrates the theory with numerical examples.

Without knowing the works of Bravais, Bienaymé, and Helmert, the Dutch mathematician C.M. Schols (1849–1897) in two papers ([232], [233]) discusses bivariate and trivariate probability distributions in terms of second-order moments. By a linear transformation of the original variables, he obtains new variables with the property that the expected value of the product of any two variables equals zero. Because the central limit theorem holds for any of the transformed variables, it follows that the asymptotic distribution of the original variables is multivariate normal. He also derives the multivariate normal by generalizing Gauss's proof for the univariate case. He indicates that the bivariate normal may be used for describing the distribution of marks in target shooting.

The American mathematician E.L. de Forest (1834–1888) generalizes the gamma distribution, which he had derived in 1882–1883, to "an unsymmetrical law of error in the position of a point in space" [90] and expresses its parameters in terms of moments. In 1885 he fits a bivariate normal distribution by means of the second-order moments to several series of observations on the errors in target shooting.

It is a curious coincidence that in the same year as Galton published *Natural Inheritance* with his empirically based theory of the bivariate normal, the French mathematician and probabilist J. Bertrand (1822–1900) published *Calcul des Probabilités* [8] containing a discussion of the bivariate normal with moment estimators of the parameters, a test for bivariate normality, and an application to target shooting.

Bertrand ([8], Chapter 9) refers to Bravais and Schols and gives a simplified version of their results for the bivariate normal. He writes

$$p(x, y) = \pi^{-1}(a^2 b^2 - c^2)^{1/2} \exp(-q(x, y)),$$

where

$$q(x, y) = a^2 x^2 + 2cxy + b^2 y^2,$$

and expresses the coefficients in terms of the moments; that is, he goes as usual at the time from (16.2) to (16.1). He notes that the equation $q(x, y) = k$ defines an ellipse with area

$$\pi k(a^2 b^2 - c^2)^{-1/2} = 2\pi k(\sigma_x^2 \sigma_y^2 - \sigma_{xy}^2)^{1/2},$$

so the probability that the point (x, y) falls between the two ellipses defined by the constants k and $k + dk$ equals $e^{-k} dk$. So far this is just a restatement of Bravais's results, but Bertrand now goes one step further by developing a test for bivariate normality.

In contradistinction to Bravais, Bertrand had a set of bivariate observations, namely the results of target shootings, that he wanted to analyze. To test for bivariate normality, he divided the plane into ten regions of equal probability by means of a series of concentric ellipses. He used the empirical moments as the estimates of the theoretical; that is, he used $\sum x_i y_i / n$ as the estimate of $E(xy)$; we denote the corresponding value of q as $\hat{q}$. From the observed positions of 1000 shots, he calculated the deviations (x_i, y_i) from the mean position, the second-order moments, and $\hat{q}(x_i, y_i)$, $i = 1, \ldots, 1000$. He compared the distribution of the $\hat{q}$s over the ten regions with a uniform distribution with expectation equal to 100 for each cell, using the binomial to test the deviation for each cell, and concluded that the hypothesis of bivariate normality cannot be rejected. (He should of course have used the multinomial instead of the binomial.)

16.2 Galton's Empirical Investigations of Regression and Correlation, 1869–1890

Francis Galton (1822–1911) studied medicine and mathematics in Cambridge and graduated in 1843. His father, a Birmington banker, died in 1844 and left him a considerable fortune so for the rest of his life he was free to pursue his many interests. He financed and carried out an expedition to Southwest Africa, which at the time was largely unexplored. He collected data from meteorological stations in many European countries for making weather maps. He had a remarkable skill for constructing mechanical contrivances, which he used for making new measuring instruments and analogue machines. To illustrate the formation and working of the symmetric binomial he constructed the quincunx, a board with rows of equidistant pins and with a funnel through which a charge of small shots was passed, each shot falling to the right or left with equal probability each time it struck a pin.

He became obsessed by measuring, counting, and graphing the phenomena he began to study in anthropology, biology, sociology, genetics, psychology, and personal identifications. He established an anthropometric laboratory for collecting measurements of the various characteristics, physical and mental, of human beings. His main interest from the mid-1860s to 1890 was empirical studies of laws of heredity by statistical methods.

Galton enriched the statistical vocabulary with several new terms. He objected to the term "error" for biological variation; instead he used "deviation." Because anthropological measurements "normally" follow the "law of errors," he rechristened this law as "the Normal curve of distributions" and wrote about "the normal deviate." Quartile, decile, and percentile are also due to him, whereas median had been used by Cournot. He introduced the terms "regression" and "correlation" in his studies of the bivariate normal distribution. When he began his statistical work in the 1860s, the methods of Laplace and Gauss and their followers were not generally known in Britain. Galton therefore developed his own crude methods, numerical and graphical, for analyzing normally distributed observations in one and two dimensions. Although his methods were primitive, his ideas were clearly expressed and had a profound effect on the development of the British Biometric School.

Galton characterizes the location of a distribution by the median M and the dispersion by the probable deviation defined as half the interquartile range $Q = \frac{1}{2}(Q_3 - Q_1)$; he finds these quantities by interpolation on the cumulative distribution. Let x be normal and denote the $100P$ percentile by x_P, $0 < P < 1$. Galton [94] writes $x_P = M + v_P Q$, so that

$$Q = (x_{P_2} - x_{P_1})/(v_{P_2} - v_{P_1})$$

and

$$M = x_{P_1} - v_{P_1} Q = x_{P_2} - v_{P_2} Q.$$

He tabulates v_P to two decimal places for $P = 0.01(0.01)0.99$. In this way he can find M and Q from any two conveniently chosen percentiles, which he reads off from the smoothed graph of the empirical cumulative frequencies. In 1899 he improves the method by plotting the cumulative frequencies on normal probability paper, so he only has to fit a straight line to the data. Among his many analyses of anthropometric data, we only relate his results [92] on the joint distribution of the heights of parents and adult children. To simplify the analysis, Galton introduces the average of the father's height and 1.08 times the mother's height, which he calls the height of the midparent. Let x denote the height of the midparent and y the height of an adult child. For each group of midparents, he finds the median $M(y|x)$ and the probable deviation $Q(y|x)$. Plotting $M(y|x)$ against x, he gets the diagram shown in Figure 16.2.1, which shows that

$$M(y|x) - M \cong \frac{2}{3}(x - M), \quad M = 68\frac{1}{4} \text{ inches.} \tag{16.6}$$

Moreover, he observes that $Q(y|x)$ is nearly independent of x and approximately equal to 1.5. Furthermore, he remarks that

$$M(x|y) - M \cong \frac{1}{3}(y - M), \tag{16.7}$$

which differs from the result obtained by solving (16.6) for x. This "apparent paradox" of the different regression lines caused Galton to seek an explanation, which he found by studying the structure of the two-way table of heights; see Figure 16.2.2.

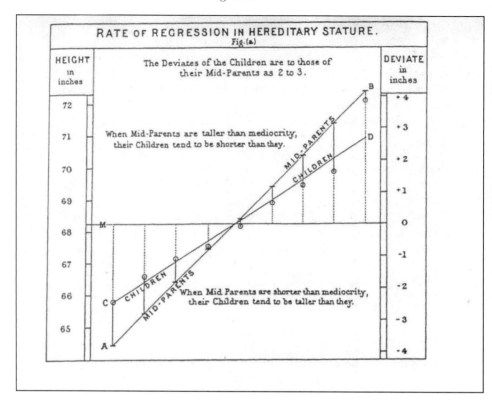

Fig. 16.2.1. Galton's [92] graph of the regression for the height of adult children on the height of midparents.

By interpreting the empirical distribution as a surface in three dimensions, by drawing lines of equal frequency, similar to isobars on a weather chart, and by smoothing these lines, Galton thus derives the basic feature of the bivariate normal distribution. By means of the series of concentric ellipses, he demonstrates how the regression lines, defined as the row and column medians, correspond to the loci of the horizontal and vertical tangential points, as shown in the diagram.

Galton gave up the mathematical derivation of the bivariate distribution; instead he "disentangled [the problem] from all reference to heredity" and submitted it to the mathematician J. Hamilton Dickson in terms of "three elementary data, supposing the law of frequency of error to be applicable throughout" ([93], p. 255). In our formulation the three data are (1) $E(x) = 0$ and $V(x) = \sigma^2$, (2) $E(y|x) = rx$, and (3) $V(y|x) = \omega^2$. Essentially Dickson wrote the distribution in the form

$$p(x, y) \propto e^{-q/2},$$

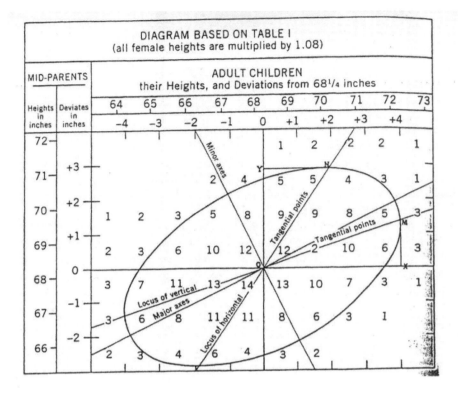

Fig. 16.2.2. Galton's [92] diagram showing the "smoothed" version of the two-way height table, the contour ellipse with its axes, the two regression lines, and the horizontal and vertical tangents.

where

$$q = \frac{x^2}{\sigma^2} + \frac{(y - rx)^2}{\omega^2} \tag{16.8}$$

$$= \frac{r^2\sigma^2 + \omega^2}{\sigma^2\omega^2}x^2 - 2\frac{r}{\omega^2}xy + \frac{1}{\omega^2}y^2$$

$$= \frac{r^2\sigma^2 + \omega^2}{\sigma^2\omega^2}\left(x - \frac{r\sigma^2}{r^2\sigma^2 + \omega^2}y\right) + \frac{1}{r^2\sigma^2 + \omega^2}y^2,$$

which shows that

$$V(y) = r^2\sigma^2 + \omega^2, \tag{16.9}$$

$$E(x|y) = \frac{r\sigma^2}{r^2\sigma^2 + \omega^2}y = r\frac{V(x)}{V(y)}y, \tag{16.10}$$

and

$$V(x|y) = \frac{\sigma^2 \omega^2}{r^2 \sigma^2 + \omega^2} = V(x)\frac{V(y|x)}{V(y)}. \tag{16.11}$$

The tangential points are found by setting the derivatives of q equal to zero. Denoting the two regression coefficients by $r_{y|x}$ and $r_{x|y}$, it follows that

$$r_{x|y}V(y) = r_{y|x}V(x); \tag{16.12}$$

see Galton ([95], p. 57). Dickson expresses the results in terms of the modulus $\sigma\sqrt{2}$, whereas Galton uses the probable deviation.

Galton provides a further argument for the agreement of the empirical distribution and the bivariate normal by demonstrating that the empirical variances and regression coefficients with good approximation satisfy the relations (16.9) and (16.12).

After having completed the proofreading of *Natural Inheritance* in the fall of 1888, it dawned upon Galton that he had overlooked an essential property of the two regression equations and that the concepts of regression and correlation, as he called it, were not limited to hereditary problems but were applicable in many other fields. He writes [96]:

> Fearing that this idea, which had become so evident to myself, would strike many others as soon as *Natural Inheritance* was published, and that I should be justly reproached for having overlooked it, I made all haste to prepare a paper, for the Royal Society with the title of "Correlation."

The full title is "Co-relations and Their Measurement, Chiefly from Anthropometric Data" [95]; in the next paper, he used "correlation" instead of "co-relation". He explains that two related problems led him to the new idea. The first is the problem of estimating the height of an unknown man from the length of a particular bone dug out of an ancient grave; the other is the problem of identifying criminals by anthropometric measurements, as proposed by A. Bertillon, specifically the value of including more bodily dimensions of the same person.

In his anthropometric laboratory he had measured the following characteristics of about 350 males: length and breath of head, stature, left middle finger, left cubit, and height of right knee, cubit being the distance between the elbow of the arm and the tip of the middle finger. In the 1889b paper he presents the regression analyses of these data and formulates his new idea as follows (p. 136):

> These relations [regressions] are not numerically reciprocal, but the exactness of the co-relation becomes established when we have transmuted the inches or other measurement of the cubit and of the stature into units dependent on their respective scales of variability. ... The particular unit that I shall employ is the value of the probable error of any single measure in its own group.

The simple idea of standardization thus united the two regressions and led to a new concept: the correlation coefficient.

Expressed in mathematical terms he transforms the relations into the co-relations

$$\frac{M(y|x) - M(y)}{Q(y)} = r_{y|x}\frac{Q(x)}{Q(y)}\frac{x - M(x)}{Q(x)} \tag{16.13}$$

and

$$\frac{M(x|y) - M(x)}{Q(x)} = r_{x|y}\frac{Q(y)}{Q(x)}\frac{y - M(y)}{Q(y)}. \tag{16.14}$$

From (16.12) it follows that

$$r_{y|x}\frac{Q(x)}{Q(y)} = r_{x|y}\frac{Q(y)}{Q(x)}, \tag{16.15}$$

so the two standardized regression lines have the same slope, which Galton ([95], p. 143) calls the index of co-relation.

Let us denote this index by R and the standardized variables by X and Y. Galton's standardized regressions can then be written as

$$M(Y|X) = RX \text{ and } M(X|Y) = RY. \tag{16.16}$$

Moreover, as noted by Galton, it follows from (16.9) that

$$Q^2(Y|X) = (1 - R^2)Q^2(Y) \text{ and } Q^2(X|Y) = (1 - R^2)Q^2(X). \tag{16.17}$$

To convince the reader of the usefulness of this method, Galton carries out 2×7 regression analyses according to formulas (16.13) and (16.14). For stature and cubit he presents the two-way table of observations from which he derives $M(y|x)$ and $M(x|y)$ by his graphical estimation procedure. Plotting the standardized deviations, he obtains the diagram shown in Figure 16.2.3, from which he reads off the common slope $R \cong 0.8$. According to (16.17) the ratio of the conditional and the marginal probable deviations equals 0.6. For the other measured characteristics he similarly determines the correlation coefficient and the conditional probable deviation.

Galton ends his paper with the remark that the same method may be used to measure the degree in which "one variable may be co-related with the combined effect of n other variables."

In his last paper on correlation [96] Galton notes that the bivariate normal distribution is determined by five parameters: the marginal medians and probable deviations and the index of correlation, which is calculated from (16.15). He explains that correlation is due to

> ... the occurrence of three independent sets of variable influences, which we have called (1), (2), and (3). The set (1) influences both events, not necessarily to the same degree; the set (2) influences one member of the pair exclusively; and the set (3) similarly influences the other member. Whenever the resultant variability of the two events is on a similar scale, the relation becomes correlation.

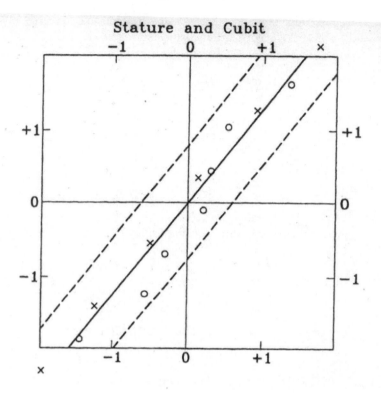

Fig. 16.2.3. Galton's [95] two standardized regressions for stature and cubit. The common slope equals 0.8. Galton calls the two dashed lines the lines of Q_1 and Q_3 values. They are drawn at the vertical distance of 0.6 from the regression line.

He illustrates this idea by several examples. The first is the correlation between dimensions of the same person, the second the correlation between the arrival times of two clerks who leave their common office by bus at the same time and then walk to their respective homes from the same halting place, and the third is the correlation between the profits of two investors who both have shares in the same company but also shares in different companies.

He remarks that "There seems to be a wide field for the application of these methods, to social problems," and adds prophetically "I can only say that there is a vast field of topics that fall under the laws of correlation, which lies quite open to the research of any competent person who cares to investigate it." This challenge was taken up by Edgeworth, Pearson, and Yule as sketched in the next section.

16.3 The Mathematization of Galton's Ideas by Edgeworth, Pearson, and Yule

Galton's epoch-making analysis of the anthropometric data, leading to the two regression lines and the correlation coefficient as characteristics of the bivariate normal distribution, blew new life into British statistical theory which for a long time had lived in the shadow of the developments on the Continent. From 1892 on, British statisticians developed a mathematical version and extension of Galton's ideas that proved so fruitful that the British school became the leading one in the further advancement of statistical theory and its applications. However, the break in the historical development was not as great as British statisticians imagined. Gradually it dawned upon them that many of the mathematical properties of the multivariate normal distribution that they discovered were to be found in the Continental literature. In particular, they did not realize that their regression analysis was a version of the linear model and the linear estimation theory. On the other hand, their belief in being pioneers and the successful applications of their methods to problems in biology, economics, and the social sciences created great enthusiasm and a drive for further development.

Edgeworth was the first statistician to take up and generalize Galton's ideas. He writes [44] $p(x) \propto \exp\{-(x - \mu)'A(x - \mu)\}$ and asks the questions: "What is the most probable value of one deviation x_r, corresponding to assigned values of x_1', x_2', etc. of the other variables?" and "What is the dispersion of the values of x_r, about its mean (the other variables being assigned)?"

He proposes to show how the elements of A are calculated from given values of ρ_{ij} and remarks:

> The problem has been solved by Mr. Galton for the case of two variables. The happy device of measuring each deviation by the corresponding quartile taken as a unit enables him to express the sought quadratic in terms of a single parameter, as thus

$$q(x_1, x_2) = \frac{x_1^2}{1 - \rho^2} - \frac{2\rho x_1 x_2}{1 - \rho^2} + \frac{x_2^2}{1 - \rho^2}; \qquad (16.18)$$

> where our ρ is Mr. Galton's r.

Hence Edgeworth follows Galton by using standardized variables, but he replaces Galton's median and probable deviation by the mean and the modulus ($\sigma\sqrt{2}$). Instead of Galton's [95] "index of co-relation" he uses "correlation coefficient" for the new parameter ρ. Edgeworth is thus the first to use the form (16.3). However, he does not define ρ as $E(x_1 x_2)$ but adheres to Galton's definition of ρ as a regression coefficient for the standardized variables.

The first problem is to answer the two questions above. Using that the mean of the normal distribution is the most probable value of the variable, he determines ξ_2, the expected value of x_2 for $x_1 = x_1'$, by solving the equation $\partial q(x_1', x_2)/\partial x_2 = 0$, which leads to $\xi_2 = \rho x_1'$. Similarly he obtains $\xi_1 = \rho x_2'$. By the usual decomposition of q, we have

$$q(x_1', x_2) = (1 - \rho^2)^{-1}(x_2 - \rho x_1')^2 + x_1'^2,$$

which shows that x_2 for given x_1 is normal with mean ρx_1 and variance $1 - \rho^2$, the reciprocal of the coefficient of x_2^2. Edgeworth's proof corresponds to Hamilton Dickson's proof in reverse order.

Writing q in the alternative form

$$q(x_1, x_2) = a_{11}x_1^2 + 2a_{12}x_1x_2 + a_{22}x_2^2, \tag{16.19}$$

the next problem is to express the coefficients in terms of ρ. Solving the equations $\partial q/\partial x_1 = 0$ and $\partial q/\partial x_2 = 0$, we get

$$\rho = -\frac{a_{12}}{a_{11}} = -\frac{a_{12}}{a_{22}},$$

and using that the coefficient of x_2^2 is the reciprocal conditional variance we get $1 - \rho^2 = a_{22}^{-1}$ and similarly $1 - \rho^2 = a_{11}^{-1}$, so $a_{12} = -\rho/(1 - \rho^2)$.

Edgeworth then considers the case of three variables assuming that the correlation coefficient for each pair is known. He refers to Galton's example, [95], where the correlations among stature, cubit, and height of knee are calculated. Setting $q(x_1, x_2, x_3) = x'Ax$, he derives $p(x_1, x_2)$ by integration with respect to x_3, and comparing the quadratic form in the exponent with (16.18) for $\rho = \rho_{12}$, he finds

$$\rho_{12} = \frac{a_{13}a_{23} - a_{12}a_{33}}{a_{22}a_{33} - a_{23}^2} = \frac{A_{12}}{A_{11}},$$

where $\{A_{ij}\}$ are the cofactors of A. By permutation of the indexes, he gets the analogous expressions for ρ_{13} and ρ_{23}. Integrating $p(x_1, x_2)$ with respect to x_2, he finds $p(x_1)$, and using the fact that x_1 has unit variance he gets $A_{11} = |A|$. Hence $\rho_{12} = A_{12}/|A|$ which means that $C = A^{-1}$.

Edgeworth indicates the solution for $m = 4$, and in a later paper [45] he gives the general solution in determinantal form. Pearson ([202], § 10b) named the multivariate normal in the form (16.3 Edgeworth's theorem.

For $m = 3$ he answers the original two questions by saying that the most probable value of x_3, say, for $x_1 = x_1'$ and $x_2 = x_2'$ is found by solving the equation $\partial q/\partial x_3 = 0$ and that the conditional variance is the reciprocal of the coefficient of x_3^2. Using this method in the general case, we get

$$E(x_m|x_1', \ldots, x_{m-1}') = a_{mm}^{-1}(a_{1m}x_1' + a_{2m}x_2' + \cdots + a_{m-1,m}x_{m-1}') \tag{16.20}$$

and

$$V(x_m|x_1', \ldots, x_{m-1}') = a_{mm}^{-1}. \tag{16.21}$$

Pearson [202] gives a clear mathematical account of the theory of normal correlation with applications to heredity and biometry; this paper must have convinced many natural scientists and mathematicians of the importance of this new tool for the analysis of biological observations.

He begins by assuming that the random variables $x_1, \ldots, x_m$ are linear functions of the n independently and normally distributed variables $\varepsilon_1, \ldots, \varepsilon_n$, $n > m$. Following Bravais, he eliminates $n - m$ of the εs from the quadratic form in the exponent of $p(\varepsilon_1, \ldots, \varepsilon_n)$, and thus he gets $p(x_1, \ldots, x_m)$ in the form (16.1) from which he derives (16.2).

For the bivariate case he introduces the correlation by means of the relation $\rho = -a_{12}/\sqrt{a_{11}a_{22}}$. Like Edgeworth he uses the same symbol for the sample and population values; Edgeworth uses ρ, whereas Pearson, like Galton, uses r. We follow Soper [240] by using r for the sample and ρ for the population value.

To estimate ρ, Pearson follows the usual procedure at the time, namely to maximize the posterior distribution. For a sample of n observations, he gets

$$p(\sigma_1, \sigma_2, \rho | \underline{x}, \underline{y}) \propto \frac{1}{(1 - \rho^2)^{n/2}} \exp\left[-\frac{1}{(1 - \rho^2)} \sum \left(\frac{x_i^2}{\sigma_1^2} - \frac{2\rho x_i y_i}{\sigma_1 \sigma_2} + \frac{y_i^2}{\sigma_2^2} \right) \right].$$
$$(16.22)$$

In the paper [213] together with L.N.G. Filon on the probable error of frequency constants, they derive the formula

$$V(r) = \frac{\left(1 - r^2\right)^2}{n}. \qquad (16.23)$$

In the 1896 paper Pearson derives the formulas for the multivariate normal regression coefficients in terms of the standard deviations and correlation coefficients, and in the 1898 paper Pearson and Filon find the standard errors of all the "frequency constants" involved. Pearson thus consolidated the large sample theory of estimation for the multivariate normal distribution and presented it in a workable form with examples from anthropometry, heredity, biology, and vital statistics.

The third step in the development of multivariate analysis in Britain was taken by G.U. Yule (1871–1951). After having studied engineering, he became assistant to Pearson from 1893 to 1899. Between 1899 and 1912 he worked as secretary for an examining body on technology in London and besides he held the Newmarch Lectureship in statistics 1902–1909. In 1911 Yule published *Introduction to the Theory of Statistics*, one of the best and most popular early textbooks in English. In 1912 he became lecturer in statistics at Cambridge University. He became known for his many original contributions to the theory and application of regression, correlation, association of attributes, the compound Poisson distribution, autoregressive time series, nonsense correlations between time series, and the study of literature style.

In the introduction to his first paper on regression, Yule [270] writes:

The only theory of correlation at present available for practical use is based on the normal law of frequency, but, unfortunately, this law is not valid in a great many cases which are both common and important. It does not hold good, to take examples from biology, for statistics of fertility in man, for measurements on flowers, or for weight measurements even on adults. In economic statistics, on the other hand, normal distributions appear to be highly exceptional: variation of wages, prices, valuations, pauperism, and so forth, are always skew. In cases like these we have at present no means of measuring the correlation by one or more "correlation coefficients" such are afforded by the normal theory.

He points out that in such cases statisticians in practice are looking for "a single-valued relation" between the variables. Suppose that the n observations of two variables (x, y) are grouped into a two-way or correlation table. Following Pearson, he calls a row or a column an array, the (conditional) distribution of the n_i observations in an array being characterized by the midpoint of the class interval, x_i say, and the mean and variance of the other variable, $\bar{y}_i$ and s_i^2, $i = 1, \ldots, k$. Yule illustrates the relation by plotting $\bar{y}_i$ against x_i and proposes to fit a straight line to the points. He makes no assumptions on the conditional distribution of y for given x and on the form of the true curve connecting y and x. In the following x and y denote deviations from their respective arithmetic means. He determines the line by the method of least squares:

> I do this solely for convenience of analysis; I do not claim for the method adopted any peculiar advantage as regards the probability of its results. It would, in fact, be absurd to do so, for I am postulating at the very outset that the curve of regression is only exceptionally a straight line; there can consequently be no meaning in seeking for the most probable straight line to represent the regression.

From the algebraic identity

$$\sum_{i=1}^{k}\sum_{j=1}^{n_i}\{y_{ij} - (a + bx_i)\}^2 = \sum_{i=1}^{k} n_i s_i^2 + \sum_{i=1}^{k} n_i d_i^2, \; d_i = \bar{y}_i - a - bx_i, \quad (16.24)$$

it follows that $\sum n_i d_i^2$ is minimized by minimizing the left side of the equation, because $\sum n_i s_i^2$ is independent of a and b. He calls this line the regression of y on x. Similarly there is a regression of x on y.

The idea may be extended to several variables. By minimizing

$$\sum [x_{1i} - (a_{12}x_{2i} + \cdots + a_{1m}x_{mi})]^2,$$

he gets the regression of x_1 on $(x_2, \ldots, x_m)$. He carries out this process for three and four variables, noting that the minimization leads to the normal equations.

Yule finds the two regression coefficients as

$$b_1 = \frac{\sum x_i y_i}{\sum y_i^2} = \frac{r s_1}{s_2} \text{ and } b_2 = \frac{\sum x_i y_i}{\sum x_i^2} = \frac{r s_2}{s_1}, \tag{16.25}$$

where $r = \sum x_i y_i / n s_1 s_2$, so that $r = \sqrt{b_1 b_2}$. Inserting these values in the sum of squares of deviations, he obtains the residual (or conditional) variances $s_1^2(1 - r^2)$ and $s_2^2(1 - r^2)$, respectively.

For three variables he writes

$$x_1 = b_{12} x_2 + b_{13} x_3,$$

and by solving the normal equations, he finds

$$b_{12} = \frac{r_{12} - r_{13} r_{23}}{1 - r_{23}^2} \frac{s_1}{s_2} \text{ and } b_{13} = \frac{r_{13} - r_{12} r_{23}}{1 - r_{23}^2} \frac{s_1}{s_3}. \tag{16.26}$$

Finally he expresses the residual variance as

$$s_1^2(1 - R_1^2),$$

where

$$R_1^2 = \frac{r_{12}^2 + r_{13}^2 - 2 r_{12} r_{13} r_{23}}{1 - r_{23}^2}, \tag{16.27}$$

and remarks that R_1, today called the multiple correlation coefficient, can be regarded as a coefficient of correlation between x_1 and (x_2, x_3).

His paper was useful for his intended audience of economists and social scientists. He showed them how to calculate "a single-valued relation" expressing the average value of a dependent variable as a linear function of independent variables. He pointed out that for economic and social data relations are often one-sided so that regression is the right tool to use, in contradistinction to anthropometric data where the organs enter symmetrically in the analysis inasmuch as one is the cause of another.

About a year later Yule [271] published a slightly revised version of the 1897a paper supplemented by two economic examples and a survey of the theory of normal correlation. In the introduction he stresses the difference between experimental data from physics and observational data from economics. He mentions that the relations, used in regression analysis should be linear in the parameters but not necessarily in the independent variables. He [272] continues the analysis of the example by introducing two more variables to account for the possibility that the changes in pauperism between 1871 and 1891 may be ascribed either to a change in the proportion of outrelief or to a common association of both variables with a change in age distribution and population size; that is, instead of the simple relation $y = a + bx_1$, he considers $y = a + b_1 x_1 + b_2 x_2 + b_3 x_3$. Having accounted for the influence of three factors, he says that there is still a certain chance of error, depending on the number of factors omitted, that may be correlated with the four variables included, but obviously the chance of error will be much smaller than before.

Finally Yule [273] gives a review of the theory of multivariate regression and correlation in terms of a new system of notation, which became widely used after he applied it in his textbook [274]. The idea may be illustrated by writing the regression equation in the form

$$x_1 = b_{12.34...n}x_2 + b_{13.24...n}x_3 + \cdots + b_{1n.23...n-1}x_n.$$

In cooperation with Engledow (Engledow and Yule, [57]), Yule also developed the method of minimum chi-squared; see Edwards [52]. It was independently discussed by Smith [238].

16.4 Orthogonal Regression. The Orthogonalization of the Linear Model

Consider the special case $y = Z\gamma + \varepsilon$ of the linear model, where $Z = (z_1, \ldots, z_m)$ consists of m linearly independent orthogonal vectors and $Z'Z = D$, D being diagonal. Denoting the least squares estimate of γ by c, the normal equations are $Z'y = Dc$, so

$$c_r = z_r'y/d_r, \ r = 1, 2, \ldots, m. \tag{16.28}$$

The dispersion matrix for the cs is $\sigma^2 D^{-1}Z'ZD^{-1} = \sigma^2 D^{-1}$, so the cs are uncorrelated. Moreover, the residual sum of squares equals

$$e'e = (y - Zc)'(y - Zc) = y'y - c_1^2 d_1 - \cdots - c_m^2 d_m. \tag{16.29}$$

Hence, for the orthogonal model the coefficients are determined successively and independently of each other and the effect of adding a new independent variable to the model is easily judged by the decrease in the residual sum of squares.

It is thus natural to ask the question: Is it possible to reparameterize the general linear model to take advantage of the simple results above?

Setting $X = ZU$, where U is an $(m \times m)$ unit upper triangular matrix we have

$$y = X\beta + \varepsilon = Z\gamma + \varepsilon, \ \gamma = U\beta, \tag{16.30}$$

where the parameter β has been replaced by γ. The problem is thus to find Z from X.

The orthogonalization of the linear model in connection with the method of least squares is due to Chebyshev and Gram. Cheybyshev's [30] proof is cumbersome and artificial; he derives the orthogonal vectors as convergents of continued fractions. Gram's [110] proof is much simpler.

From $X = (x_1, \ldots, x_m)$ we get the matrix of coefficients for the normal equations as $A = \{a_{rs}\}$, $a_{rs} = x_r'x_s$, $(r, s) = 1, \ldots, m$. Gram introduces z_r as the linear combination of $x_1, \ldots, x_r$ defined as $z_1 = x_1$ and

$$z_r = \begin{vmatrix} a_{11} & \cdots & a_{1r} \\ \vdots & & \vdots \\ a_{r-1,1} & \cdots & a_{r-1,r} \\ x_1 & \cdots & x_r \end{vmatrix}, \quad r = 2, \ldots, m. \qquad (16.31)$$

It follows that

$$z_r = \sum_{s=1}^{r} A_{rs}^{(r)} x_s,$$

where $A_{rs}^{(m)}$ is the cofactor of a_{rs} in the determinant $A^{(m)} = |a_{rs}|$. The orthogonality of the zs follows from the fact that

$$x_k' z_r = \sum_{s=1}^{r} A_{rs}^{(r)} a_{ks} = 0, \text{ for } k < r.$$

The least squares estimation of β may then be carried out in two steps. First, γ is estimated by $c = D^{-1} Z' y$ and next β is estimated by $b = U^{-1} c$. The price to be paid for this simple solution is the calculation of Z but this is an easy matter for $m \le 4$, say.

Gram has thus proved that the adjusted value of y by the method of least squares when m terms of the linear model are included, that is,

$$\hat{y}^{(m)} = x_1 b_{m1} + \cdots + x_m b_{mm},$$

may be transformed to

$$\hat{y}^{(m)} = z_1 c_1 + \cdots + z_m c_m. \qquad (16.32)$$

Writing

$$\hat{y}^{(m)} = \hat{y}^{(1)} + (\hat{y}^{(2)} - \hat{y}^{(1)}) + \cdots + (\hat{y}^{(m)} - \hat{y}^{(m-1)}), \qquad (16.33)$$

and noting that

$$\hat{y}^{(m)} - \hat{y}^{(m-1)} = z_m c_m,$$

Gram gets the fundamental result that the successive terms of the decomposition (16.33) are orthogonal. Gram's decomposition expresses the fact that the explanatory variable x_1 leads to the adjusted value $\hat{y}^{(1)}$, and the two explanatory variables (x_1, x_2) lead to $\hat{y}^{(2)}$, so the net effect of taking x_2 into account is $\hat{y}^{(2)} - \hat{y}^{(1)}$, which is orthogonal to $\hat{y}^{(1)}$, and so on.

Sampling Distributions Under Normality, 1876–1908

17.1 The Distribution of the Arithmetic Mean

In the present chapter it is assumed that $x_1, \ldots, x_n$ are independently and normally distributed (μ, σ^2).

Gauss [100] proved that the arithmetic mean $\bar{x}$ is normally distributed $(\mu, \sigma^2/n)$, assuming that the prior distribution of μ is uniform.

Laplace [157] proved the central limit theorem, which implies that $\bar{x}$ is asymptotically normal $(\mu, \sigma^2/n)$, with the frequentist interpretation of probability. As far as we know he did not give a proof for finite n but a proof is easily constructed from his general methods of analysis. From the convolution formula for the distribution of the sum of two independent random variables it follows that $x_1 + x_2$ is normal $(2\mu, 2\sigma^2)$ and by iteration the general result is obtained. From his characteristic function for the normal distribution $\psi(t) = \exp(i\mu t - \sigma^2 t^2/2)$ it is easy to find $\psi^n(t)$ and by the inversion formula to find the distribution of $\bar{x}$.

A simple direct proof is given by Encke ([56], p. 278). He writes

$$p(x_1, \ldots, x_n) = \pi^{-n/2} h^n \exp\{-h^2 \sum (x_i - \bar{x})^2 - h^2 (\bar{x} - \mu)^2 n\},$$

from which he concludes that

$$p(\bar{x}) = (n\pi)^{-1/2} h \exp(-h^2 (\bar{x} - \mu)^2 n).$$

This became the textbook method of proving the normality of $\bar{x}$.

17.2 The Distribution of the Variance and the Mean Deviation by Helmert, 1876

Friedrich Robert Helmert (1843–1917) got his doctor's degree from the University of Leipzig in 1867 for a thesis of higher geodesy. In 1870 he became instructor and in 1872 professor of geodesy at the Technical University in Aachen. From 1887 he was professor of geodesy at the University of Berlin and director of the Geodetic Institute. He became famous for his work on the mathematical and physical theories of higher geodesy and for his book on the adjustment of observations *Die Ausgleichungsrechnung nach der Methode der kleinsten Quadrate mit Anwendungen auf die Geodäsie und die Theorie der Messinstrumente* [119]. He supplemented this book with several papers on error theory and included most of his results in the much enlarged second edition, 1907. Writing for geodesists, he did not explain the basic principles of statistical inference but kept to the practical aspects of the method of least squares according to Gauss's second proof, although he also mentioned the first. His book is a pedagogical masterpiece; it became a standard text until it was superseded by expositions using matrix algebra.

Helmert [121] derives the distribution of $[\varepsilon\varepsilon]$ by induction. Let $y = \varepsilon_1^2$. Then

$$p_1(y)dy = \int f(\varepsilon)d(\varepsilon) \text{ for } y \le \varepsilon^2 \le y + dy \tag{17.1}$$

$$= 2 \int f(\varepsilon)d\varepsilon \text{ for } y^{1/2} \le \varepsilon \le y^{1/2} + \frac{1}{2}y^{-1/2}dy$$

$$= (2\pi\sigma^2)^{-1/2}y^{-1/2}\exp(-\frac{y}{2\sigma^2})dy.$$

Set $y = \varepsilon_1^2 + \varepsilon_2^2 = y_1 + y_2$. Then

$$p_2(y) = \int_0^y p_1(y)p_1(y - y_1)dy_1 \tag{17.2}$$

$$= (2\pi\sigma^2)^{-1}\exp\left(-\frac{y}{2\sigma^2}\right)\int_0^y y_1^{-1/2}(y - y_1)^{-1/2}dy_1$$

$$= (2\sigma^2)^{-1}\exp\left(-\frac{y}{2\sigma^2}\right).$$

Combining (17.1) and (17.2), and (17.2) with itself, he gets the distributions for sums of three and four components. The distribution of $y = [\varepsilon\varepsilon]$ is defined as

$$p_n(y)dy = (2\pi\sigma^2)^{-n/2}\underset{y\le[\varepsilon\varepsilon]\le y+dy}{\int\cdots\int}\exp\left(-\frac{[\varepsilon\varepsilon]}{2\sigma^2}\right)d\varepsilon_1\cdots d\varepsilon_n. \tag{17.3}$$

Helmert states that

$$p_n(y) = \frac{1}{2^{n/2}\Gamma\left(\frac{1}{2}n\right)\sigma^n}y^{(n/2)-1}\exp\left(-\frac{y}{2\sigma^2}\right). \tag{17.4}$$

To prove this by induction he sets $z = \varepsilon_{n+1}^2 + \varepsilon_{n+2}^2$ and $v = y + z$. Evaluating the integral

$$p(v) = \int_0^v p_n(y)p_2(v-y)dy,$$

he finds that $p(v) = p_{n+2}(v)$, which completes the proof.

Because the true errors usually are unknown, Helmert [122] proceeds to study the distribution of $[ee]$ where

$$\varepsilon_i = e_i + \bar{\varepsilon}, \ i = 1, \ldots, n, \ \bar{\varepsilon} = \frac{[\varepsilon]}{n}, \ [e] = 0. \tag{17.5}$$

Introducing the new variables into $p(\underline{\varepsilon})$, and using that the Jacobian of the transformation (17.5) equals n, he obtains

$$p(e_1, \ldots, e_{n-1}, \bar{\varepsilon}) = n(2\pi\sigma^2)^{-n/2} \exp\left\{ -\frac{1}{2}\sigma^{-2}([ee] + n\bar{\varepsilon}^2) \right\}, \tag{17.6}$$

which shows that the mean is independent of the residuals and that $\bar{\varepsilon}$ is normal $(0, \sigma^2/n)$. However, Helmert does not make this remark; for him $\bar{\varepsilon}$ is a variable that has to be removed by integration so that

$$p(e_1, \ldots, e_{n-1}) = n^{1/2}(2\pi\sigma^2)^{-(n-1)/2}\exp(-[ee]/2\sigma^2). \tag{17.7}$$

To find the distribution of $x = [ee]$, Helmert used the transformation

$$t_i = \sqrt{\frac{i+1}{i}} \left(e_i + \frac{1}{i+1}e_{i+1} + \cdots + \frac{1}{i+1}e_{n-1} \right), \ i = 1, \ldots, n-2, \tag{17.8}$$

$$t_{n-1} = \sqrt{\frac{n}{n-1}}e_{n-1},$$

with the Jacobian $\sqrt{n}$. Because

$$[tt] = \sum_{i=1}^{n-1} t_i^2 = \sum_{i=1}^n e_i^2 = [ee],$$

he gets from (17.7) that

$$p(x)dx = (2\pi\sigma^2)^{-(n-1)/2} \int \cdots \int_{x \leq [tt] \leq x+dx} \exp\left(-\frac{[tt]}{2\sigma^2}\right) dt_1 \cdots dt_{n-1}.$$

Comparing with (17.3), he concludes that the distribution of $[ee]$ is the same as the distribution of the sum of squares of $n-1$ true errors; that is, $p(x) = p_{n-1}(x)$ as given by (17.4).

Helmert used $s^2 = [ee]/(n-1)$ as an unbiased estimate of σ^2; he was the first to derive its distribution. Today the Helmert distribution $p(\bar{x}, s^2)$ is highly valued as the starting point for modern small-sample theory. Helmert did not realize its importance; he did not even mention the distribution in the second edition of his book, 1907.

Helmert uses $p(s^2)$ to find the mean and mean square error of s,

$$E(s) = \sigma\Gamma\left(\frac{n}{2}\right)\sqrt{2/(n-1)}/\Gamma\left(\frac{n-1}{2}\right) \text{ and } E\left(s-\sigma\right)^2 = 2\sigma^2(1 - E\left(s/\sigma\right)).$$
(17.9)

Hence, s is a biased estimate of σ, the bias being of order n^{-1}, and for large n the mean square error and thus the variance of s equals $\sigma^2/2(n-1)$. The large-sample results had previously been derived by Gauss [102].

Helmert did not know how to use the skew distribution of s to find asymmetric confidence limits for σ; he kept to the classical symmetrical large-sample result.

The mean (absolute) deviation is defined as $\sum|e_i|/n$. The German astronomer C.A.F. Peters [215] introduced

$$\frac{\sum|e_i|}{n}\sqrt{\frac{\pi n}{2(n-1)}}$$
(17.10)

as an unbiased estimate of σ. This is called Peters's formula.

Helmert [122] considers the modified statistics $m = \sum|e_i|/\sqrt{n(n-1)}$. By suitable and cumbersome transformations of $p(\varepsilon)$ he obtains $E(m) = \sigma\sqrt{2/\pi}$ and

$$V(m) = 2\sigma^2(\pi n)^{-1}\left\{\pi/2 + \sqrt{n(n-2)} - n + \arcsin(n-1)^{-1}\right\}$$
(17.11)
$$= \sigma^2(\pi n)^{-1}\left\{\pi - 2 + (2n)^{-1} + O(n^{-2})\right\}.$$

As a third estimate of σ, Helmert [122] considers the mean absolute difference

$$\bar{d} = \Sigma\Sigma\left|x_i - x_j\right|/n(n-1) = \Sigma\Sigma\left|\varepsilon_i - \varepsilon_j\right|/n(n-1), \ i < j,$$

and proves that $r = \bar{d}\sqrt{\pi}$ is an unbiased estimate of σ with variance

$$V(r) = \frac{\pi\sigma^2}{n(n-1)}\left(\frac{n+1}{3} + \frac{2(n-2)\sqrt{3} - 4n + 6}{\pi}\right).$$
(17.12)

Finally, he finds the relative efficiency of the three estimates by comparing the three standard errors calculated from (17.10), (17.11), and (17.12) for $n = 2, 3, 4, 5, 10$. For $n > 10$ he uses the corresponding large-sample formulas in the form

$$0.707/\sqrt{n-1}, \ 0.756/\sqrt{n-1}, \ 0.715/\sqrt{n-1}.$$

Hence, by this important paper Helmert extends and completes the investigation of the relative efficiency of estimates of σ initiated by Gauss [102].

Setting $y = [ee] = (n-1)s^2$ it can be seen that Helmert's formula (17.4) implies that $(n-1)s^2$ is distributed as $\sigma^2\chi^2$ with $f = n-1$ degrees of freedom, where the χ^2 distribution is defined as

$$p(\chi^2)d(\chi^2) = \frac{1}{2^{f/2}\Gamma(f/2)}(\chi^2)^{(f/2)-1}\exp(-\chi^2/2)d(\chi^2).$$
(17.13)

17.3 Pizzetti's Orthonormal Decomposition of the Sum of Squared Errors in the Linear-Normal Model, 1892

P. Pizzetti (1960–1918), professor of geodesy and astronomy at Pisa, derives the distribution of $\varepsilon'\varepsilon$ and $e'e$. Beginning with $p(\underline{\varepsilon})$ and using the transformation

$$y = \varepsilon'\varepsilon \text{ and } x_i = \varepsilon_i y^{-1/2}, \ i = 1, \ldots, n-1,$$

with the Jacobian

$$y^{(n/2)-1}(1 - x_1^2 - \cdots - x_{n-1}^2)^{-1/2},$$

he obtains

$$p(y, x_1, \ldots, x_{n-1}) = \left(\pi^{-1/2}h\right)^n \exp(-h^2 y)y^{(n/2)-1}(1 - x_1^2 - \cdots - x_{n-1}^2)^{-1/2},$$
(17.14)

from which it follows that y is distributed as $\sigma^2\chi^2$ with n degrees of freedom. He refers to Helmert's proof of this result.

To find the distribution of $e'e$, he uses the orthogonal transformation

$$t_i = \{i(i+1)\}^{-1/2}(\varepsilon_1 + \cdots + \varepsilon_i - i\varepsilon_{i+1}), \ i = 1, \ldots, n-1, \quad (17.15)$$

and

$$v = \varepsilon_1 + \cdots + \varepsilon_n,$$

so that $t't = \varepsilon'\varepsilon - v^2/n = e'e$. Hence,

$$p(t_1, \ldots, t_{n-1}) = \left(\pi^{-1/2}h\right)^n n^{1/2} \exp\left\{-h^2(t't + v^2/n)\right\}.$$

The t's and v are thus independent and

$$p(t_1, \ldots, t_{n-1}, v) = \left(\pi^{-1/2}h\right)^{n-1} \exp(-h^2 t't). \quad (17.16)$$

It follows, as in the proof of (17.14), that $t't$ is distributed as $\sigma^2\chi^2$ with $n-1$ degrees of freedom. Because the ts are independent so are the t^2s which implies the additivity of the χ^2 distribution.

Helmert used a nonorthogonal transformation to derive the distribution of $e'e$; nevertheless Pizzetti's orthogonal transformation (17.15) is often called Helmert's transformation.

Pizzetti generalizes his proof to the linear normal model with $m < n$ parameters. His main tool is the Gaussian decomposition

$$\varepsilon'\varepsilon = e'e + (b - \beta)'X'X(b - \beta) = e'e + w'w, \quad (17.17)$$

where the elements of $w' = (w_1, \ldots, w_m)$ are independently and normally distributed with zero mean and precision h. Because $X'e = 0$, $e'e$ is a quadratic form in $e_1, \ldots, e_{n-m}$, which may be transformed to a sum of squares, $t't$ say, where the elements of $t' = (t_1, \ldots, t_{n-m})$ are linear functions of $e_1, \ldots, e_{n-m}$ and thus of $\varepsilon_1, \ldots, \varepsilon_n$. Furthermore, the elements of w are linear functions of the elements of ε because $X'X(b - \beta) = X'\varepsilon$. Hence, (t', w') can be expressed as a linear transformation of ε, $\varepsilon'R'$ say, R being an $n \times n$ matrix, so that $t't + w'w = \varepsilon'R'R\varepsilon$. Because $t't = e'e$, it follows from (17.17) that $R'R = I_n$; that is, the transformation is orthonormal.

Introducing the new variables into $p(\underline{\varepsilon})$, Pizzetti finds

$$p(t_1, \ldots, t_{n-m}, w_1, \ldots, w_m) = (\pi^{-1/2}h)^n \exp(-h^2 t't - h^2 w'w). \quad (17.18)$$

This remarkable result shows that the ts are independent of the ws, and that $\varepsilon'\varepsilon$, which is distributed as $\sigma^2\chi^2$ with n degrees of freedom, has been decomposed into two independent sums of squares, $t't$ and $w'w$, that are distributed as $\sigma^2\chi^2$ with $n - m$ and m degrees of freedom, respectively. This is the theoretical basis for the analysis of variance in the fixed effects model.

Pizzetti also shows how to estimate the components of variance in the random effects model and demonstrates his theory by an example.

Finally, Pizzetti uses the distribution of $s/\sigma = \chi/\sqrt{n-m}$ to find exact confidence limits for σ, an unsolved problem at the time. Integrating the density of $\chi/\sqrt{n-m}$ for $n - m = 1, \ldots, 6$ he finds

$$P(1 - a < s/\sigma < 1 + a) \text{ for } a = 0.10, \ 0.25, \ 0.50, \ 0.75, \quad (17.19)$$

and solving for σ, he gets the confidence interval. His table is presumably the first table of the χ distribution. For $m - n > 6$ he uses the normal approximation.

17.4 Student's t Distribution by Gosset, 1908

W.S. Gosset (1876–1939) studied mathematics and chemistry at Oxford 1895–1899, whereafter he, for the rest of his life, was employed by the Guinness Brewery. In the academic year 1906–1907 he studied statistics at Pearson's Biometric Laboratory. In this environment, dominated by large-sample methods, he pursued his own problems on small-sample statistics and succeeded in deriving exact confidence limits for the population mean of normally distributed observations depending only on the observed mean and standard deviation.

The classical confidence interval for the population mean is $\bar{x} \pm us/\sqrt{n}$, where the confidence coefficient is found from the normal probability integral with argument u. This probability is only approximate because s has been used as a substitute for σ. Gosset's [254] idea is to find the sampling distribution of

$$z = \frac{\bar{x} - \mu}{s}, \quad s^2 = \sum_{i=1}^{n} (x_i - \bar{x})^2/n. \quad (17.20)$$

The confidence interval for μ then becomes $\bar{x}\pm zs$, where the probability integral for z depends on n but is independent of σ. This idea is not new. Lüroth [171] and Edgeworth [43] had solved the problem by inverse probability, assuming that h is uniformly distributed; see Section 10.2. Gosset did not know these papers and solved the problem by direct probability.

Gosset begins by noting that the distribution of $\bar{x}$, $p_1(\bar{x})$ say, is normal $(0, \sigma^2/n)$, where for convenience he sets $\mu = 0$. Next, he finds the first four moments of s^2 using that

$$s^2 = n^{-1}\Sigma x_i^2 - n^{-2}\Sigma x_i^2 - 2n^{-2}\Sigma_{i<j}x_i x_j.$$

From the powers of s^2 he finds the moments by taking expectations, and from the relation between β_1 and β_2 he guesses that the distribution of s^2 is a Pearson Type III, which he writes as

$$cs^{n-3}\exp(-s^2/2\sigma^2).$$

From the distribution of s^2 he gets

$$p_2(s) = \frac{n^{(n-1)/2}}{\Gamma\left(\frac{1}{2}(n-1)\right)2^{(n-3)/2}}\frac{s^{n-2}}{\sigma^{n-1}}e^{-ns^2/2\sigma^2}. \tag{17.21}$$

To find the correlation between $\bar{x}$ and s, he writes $\bar{x}^2 s^2$ as a symmetric function of the xs, and evaluating the expectation, he proves that there is no correlation between $\bar{x}^2$ and s^2. He uses this result as if he had proved that $\bar{x}$ and s are independent.

After these preparations he notes that $p(\bar{x}, s) = p_1(\bar{x})p_2(s)$ so that

$$p(z, s) = p_1(sz)sp_2(s). \tag{17.22}$$

Integrating with respect to s, he obtains

$$p(z) = \frac{\Gamma\left(\frac{1}{2}n\right)}{\Gamma\left(\frac{1}{2}\right)\Gamma\left(\frac{1}{2}(n-1)\right)}(1+z^2)^{-n/2}, \quad -\infty < z < \infty, \ n = 2, 3, \ldots. \tag{17.23}$$

Hence,

> Since this equation is independent of σ it will give the distribution of the distance of the mean of a sample from the mean of the population expressed in terms of the standard deviation of the sample for any normal population.

He tabulates the corresponding probability integral. Fisher suggested the transformation $t = z\sqrt{n-1}$ because t is asymptotically normal $(0,1)$. In this way Gosset's result came to be known as Student's t distribution.

For $n = 2$ (the Cauchy distribution) Gosset observes that the standard deviation is infinite, whereas the probable error is finite because $P(-1 < z < 1) = \frac{1}{2}$; that is,

...if two observations have been made and we have no other information, it is an even chance that the mean of the (normal) population
will lie between them.

He gives three examples of the usefulness of the t distribution in the analysis of data.

As noted by Gosset himself, his proof is mathematically incomplete. If he
had known Helmert's [121] paper, he could just have quoted $p_1(\bar{x})$ and $p_2(s)$
and then derived $p(z)$ by integration of (17.22). It is surprising that Pearson
did not know Helmert's result in view of the fact that he [203] refers to Czuber
[36]. It is equally surprising that Gosset did not read Thiele's [260] *Theory of
Observations*, the most advanced text in English at the time. There he could
have found the first four cumulants of s^2 for an arbitrary distribution and a
simple formula for finding all the cumulants of s^2 for the normal distribution.

Part V

THE FISHERIAN REVOLUTION, 1912–1935

Fisher's Early Papers, 1912–1921

18.1 Biography of Fisher

Ronald Aylmer Fisher (1890–1962) was born in London as the son of a prosperous auctioneer whose business collapsed in 1906, whereafter Ronald had to fend for himself. In 1909 he won a scholarship in mathematics to Cambridge University where he graduated in 1912 as a Wrangler in the Mathematical Tripos. He was awarded a studentship and spent another year in Cambridge studying statistical mechanics and quantum theory. In addition he used much of his time studying Darwin's evolutionary theory, Galton's eugenics, and Pearson's biometrical work.

Already in 1911 he demonstrated his extraordinary insight in the two scientific fields that in equal measure came to occupy him for the rest of his life: eugenics and statistics. As chairman for the undergraduate section of the Cambridge University Eugenics Society, he addressed a group of students on "Mendelism and Biometry" [61] in which he as the first indicated the synthesis of the two topics. About the same time he conceived the theory of maximum likelihood estimation published in his 1912 paper.

Leaving Cambridge in 1913, he got a statistical job with an investment company in London. When the war came he volunteered for military service but was rejected because of his poor eyesight. He spent the years 1915–1919 teaching mathematics and physics at public schools, the last two years at Bradford College in Kent. In 1917 he married Ruth Eileen Guiness. They leased a gamekeeper's cottage with adjoining land in the vicinity of the College and started subsistence farming, the daily work being carried out by Eileen and her sister.

Leonard Darwin, a younger son of Charles Darwin, was honorary president of the Eugenics Education Society and became interested in Fisher's work. The two men became friends, and Darwin supported Fisher morally and scientifically throughout his career. Between 1914 and 1934 Fisher contributed more than 200 reviews to the *Eugenics Review*.

After the war Fisher began looking for another job. He applied for a job at Cairo University but was turned down. However, in the summer of 1919 he received two offers: a temporary position as statistician at Rothamsted [Agricultural] Experimental Station and a position at Pearson's Laboratory in University College, London. He chose Rothamsted where he, during the next 15 years, developed a world-known Department of Statistics. Many British and some foreign statisticians were there introduced to Fisherian statistics as members of the staff or as visitors.

When Karl Pearson retired in 1933, his department was divided into two independent units, a Department of Statistics with Egon S. Pearson as head, and a Department of Eugenics with Fisher as Galton Professor of Eugenics, the teaching of statistics belonging to the former department.

In 1943 Fisher became professor of genetics at Cambridge where he stayed until his retirement in 1957. He spent his last three years as a research fellow in Adelaide, Australia.

A detailed description and analysis of Fisher's life and scientific work in genetics and statistics is given by his daughter Joan Fisher Box [21]. A critical review of this book is due to Kruskal [143].

The Collected Papers of R.A. Fisher have been edited by J.H. Bennett (1871–1874). Volume 1 contains a biography written by F. Yates and K. Mather [269] and a bibliography.

A survey of Fisher's contributions to statistics is given in the posthumously published paper "On rereading R.A. Fisher" by L.J. Savage [230], edited by J.W. Pratt, followed by a discussion. Another survey is due to C.R. Rao [228].

Savage ([230], § 2.3) writes that "Fisher burned even more than the rest of us, it seems to me, to be original, right, important, famous, and respected. And in enormous measure, he achieved all of that, though never enough to bring him peace."

Fisher could be polemical and arrogant. He quarrelled with Karl Pearson on the distribution of the correlation coefficient, the number of degrees of freedom for the χ^2 test, and the efficiency of the method of moments; with Gosset and others on random versus systematic arrangements of experiments; with Neyman on fiducial limits versus confidence limits; with members of the Neyman–Pearson school on the theory of testing statistical hypotheses, and with many others on specific problems.

Fisher's main work in genetics is the book *The Genetical Theory of Natural Selection* [81]. A review of Fisher's genetical work in an historical setting is due to Karlin [134].

Fisher's first and most important book is *Statistical Methods for Research Workers*, SMRW [73]. In each new edition Fisher introduced new results by adding subsections to the original one, so that the posthumously published fourteenth edition (1970) contains 362 pages compared with the original 239 pages. It has been translated into many languages.

The first edition of SMRW is a collection of prescriptions for carrying out statistical analyses of biological and agricultural data by means of the methods developed by Fisher between 1915 and 1925. It is nonmathematical and achieves its aim by using examples to demonstrate the methods of analysis and the necessary calculations. It furnishes tables of percentage points of the t, χ^2, and $z = \frac{1}{2} \ln F$ distributions for carrying out the corresponding test of significance. In the preface Fisher writes:

> The elaborate mechanism built on the theory of infinitely large samples is not accurate enough for simple laboratory data. Only by systematically tackling small sample problems on their merits does it seem possible to apply accurate tests to practical data. Such at least has been the aim of this book.

As Fisher indicates it is a great progress that the sample size now can be taken explicitly into account in tests of significance for normally distributed observations. On the other hand, he also notes that departures from normality, unless very strongly marked, can only be detected in large samples. To get out of this dilemma he appeals to the central limit theorem. In the introduction to the applications of the t test he writes (§ 23) that

> ...even if the original distribution were not exactly normal, that of the mean usually tends to normality, as the size of the sample is increased; the method is therefore applied widely and legitimately to cases in which we have not sufficient evidence to assert that the original distribution was normal, but in which we have reason to think that it does not belong to the exceptional class of distributions for which the distribution of the mean does not tend to normality.

Hence, in application of Fisher's tests, the percentage points tabulated should only be considered as approximations.

The revolutionary content of SMRW becomes evident by comparison with the many other textbooks appearing about 1925.

In the last section of SMRW Fisher discusses and exemplifies the design of agricultural experiments, pointing out that randomization, restrictions on random arrangements, such as randomized blocks and Latin squares, and replications are necessary for achieving a valid estimate of the experimental error, for the partial elimination of fertility gradients, and for increasing the sensitivity of the experiment. He shows how the total variation of the experimental results can be broken down into its constituents due to treatments, restrictions, and error by means of the analysis of variance.

In the paper "The Arrangement of Field Experiments" [77] he further classifies these principles and adds a section of "Complex Experimentation" in which he stresses the advantages of factorial experiments compared with single-factor experiments. He describes a $3 \times 2 \times 2$ factorial experiment in eight randomized blocks, each containing 12 plots, and notes as the most important advantage that the average effect of any factor by this arrangement is given "a very much wider inductive basis" than could be obtained by single-factor

experiments without extensive repetitions. Finally he indicates the possibility
of reducing the size of a multifactor experiment by confounding. He writes:

> In the above instance no possible interaction of the factors is disre-
> garded; in other cases it will sometimes be advantageous deliberately
> to sacrifice all possibility of obtaining information on some points,
> these being believed confidently to be unimportant, and thus to in-
> crease the accuracy attainable on questions of greater moment. The
> comparisons to be sacrificed will be deliberately confounded with cer-
> tain elements of the soil heterogeneity, and with them eliminated.

He presented the results of his long experience with agricultural field experiments in his second great book, *The Design of Experiments* [85], which
has exerted a profound influence on the planning of comparative experiments
not only in agriculture but in many other fields such as biology, industry,
psychology, and clinical trials. He underlines that the design and analysis of
experiments are part of a single process of the improvement of natural knowledge.

Tests of significance are dominating in Fisher's examples in both SMRW
and the *Design of Experiments*. To facilitate the application of the analysis of
variance he [73] had tabulated the 5 percentage points of the z distribution,
later supplemented with the 1 and 0.1 percentage points. This had the effect
that many research workers used these fixed p-values, although Fisher [86]
later warned against such rigid rules. He pointed out that the null hypothesis
can never be proved, but is possibly disproved, and added ([86] § 8) that
"Every experiment may be said to exist only in order to give the facts a chance
of disproving the null hypothesis." However, the meaning is that estimation
should be preceded by a test of significance, as stated in SMRW, § 51, and
shown in some of his examples.

Randomization produces a symmetric nonnormal distribution of experimental errors because part of the systematic variation (fertility gradient) is
combined with the measurement error. One might thus fear that only large-
sample theory could be applied to randomized experiments, but considering a
numerical example with 15 matched pairs of observations, Fisher shows that
the distribution of t values based on the 2^{15} randomized numbers does not
deviate much from the t distribution under normality. Without further proof
he proceeds to use the F test for randomized trials as if the error distribution
were normal, leaving the proof to his followers.

The third book, written together with F. Yates, is *Statistical Tables for
Biological, Agricultural and Medical Research* [88], which in addition to the ta-
bles contains an introduction with explanations and examples of applications.
It became an indispensable tool for the applied statistician.

The methods presented in the three books have by and large been accepted as the foundation for modern statistics in the natural sciences. The
books contain no proofs, and Fisher did not help the reader by giving explicit
references to his papers on mathematical statistics; he just listed all his papers

in chronological order at the end of SMRW. An immense literature has grown up in an attempt to explain, popularize, prove, and extend Fisher's results.

Fisher's fourth book *Statistical Methods and Scientific Inference*, SMSI [86] is an attempt to give "a rational account of the process of scientific inference as a means of understanding the real world, in the sense in which this term is understood by experimental investigators" (§ 2.1). For an understanding of Fisher's work and, in particular, his polemics, it is important to keep in mind that he always argued from the (narrow) point of view of research in the experimental sciences. Of course he recognized that statistical methods are useful for "technological, commercial, educational and administrative purposes" (see his Foreword); he even (§ 4.1) remarked that "In various ways what are known as acceptance procedures are of great importance in the modern world," at the same time warning not to confuse the logic behind tests for acceptance with the logic behind tests of significance.

The book contains a renewed discussion of the many concepts he had used in his previous work, but he does not in detail contrast his own concepts with those of the competing schools. He points out that there are many forms of statistical inference, each of them appropriate for answering specific questions under a given statistical model. They are: Bayes's theorem, tests of significance, mathematical likelihood, fiducial probability, estimation criteria, and amount of information.

Fisher was a genius who almost single-handedly created the foundation for modern statistical science without detailed study of his predecessors. When young he was ignorant not only of the Continental contributions but even of contemporary publications in English.

It is of course impossible to give a review of Fisher's monumental work in the following few pages so we limit ourselves to some selected topics.

18.2 Fisher's "Absolute Criterion," 1912

Fisher's first paper [62] on mathematical statistics, entitled "On an Absolute Criterion for Fitting Frequency Curves," was written in his third year as an undergraduate at Cambridge. Presumably he had taken a course in statistics comprising error theory with the method of least squares and Pearson curves with the method of moments. The paper begins with a rejection of these two methods of estimation and ends with proposing an absolute criterion, which he later called the method of maximum likelihood.

The paper shows a self-reliance that is remarkable in view of the fact that he was ignorant of the relevant literature. He writes under the misapprehension that the method of least squares consists of minimizing the sum of squared deviations between the observed and the true values although Gauss had defined the method in terms of the standardized deviations. Moreover, in the case of correlated observations, as in Fisher's case, it is the appropriate quadratic form that should be minimized.

He remarks correctly that the method of least squares (in his version) is inapplicable to frequency curves because the result depends on the scaling of the independent variable. However, if he had used the method correctly he would have been led to the minimization of Pearson's χ^2.

Next, he criticizes the method of moments for being arbitrary and for not giving a rule for choosing what moments to use in the estimating equations.

Fisher continues:

> But we may solve the real problem directly. If f is an ordinate of the theoretical curve of unit area, then $p = f\delta x$ is the chance of an observation falling within the range δx; and if
>
> $$\log P' = \sum_1^n \log p,$$
>
> then P' is proportional to the chance of a given set of observations occurring. The factors δx are independent of the theoretical curve, so the probability of any particular set of θ's is proportional to P, where $\log P = \sum_1^n \log f$.
>
> The most probable set of values for the θ's will make P a maximum.

Hence, $P' \propto p(\underline{x}|\underline{\theta})d\underline{x}$, the proportionality constant being $n!$, and $P = p(\underline{x}|\underline{\theta})$.

With his background in error theory Fisher naturally uses "probability" in the same sense as the astronomers; the similarity with Hagen's formulation is striking. However, for statisticians conversant with inverse probability "the probability of any particular set of θ's is proportional to P'" means that $p(\theta|x) \propto p(x|\theta)$," so that such readers get the impression that Fisher is using inverse probability. Fisher's formulation is thus ambiguous. However, in § 6, the last section of the paper, Fisher frankly declares:

> We have now obtained an absolute criterion for finding the relative probabilities of different sets of values for the elements of a probability system of known form. [...] P is a relative probability only, suitable to compare point with point, but incapable of being interpreted as a probability distribution over a region, or giving any estimate of absolute probability.

Hence, Fisher says that P is not a posterior density but the density of the observations considered as a function of the parameter. Moreover, he points out that estimates obtained by inverse probability are not invariant to parameter transformations, because the density of the transformed parameters equals the original density times the Jacobian of the transformation, whereas "the relative values of P" are invariant.

These explanations and the change of terminology regarding P means that the reader has to reread the paper, replacing inverse probability by relative probability. One wonders that the paper was published in this form; the editor should of course have demanded a revised edition incorporating the contents of § 6 in the body of the paper.

Fisher's paper did not have any impact on the statistical literature at the time. The reason for discussing it today is the fact that it contains the germ of Fisher's method of maximum likelihood. It is a weakness of the paper that it introduces a new method of estimation without indicating the properties of the resulting estimates. The solution of this problem had to wait for Fisher's 1922a paper.

18.3 The Distribution of the Correlation Coefficient, 1915, Its Transform, 1921, with Remarks on Later Results on Partial and Multiple Correlation

Twenty years after Galton conceived the idea, the correlation coefficient had found wide applications not only in biometry but also in statistical economics and experimental psychology. However, only rather few results on the properties of r were known. Pearson and Filon [213] and Sheppard [235] had proved that the large-sample standard deviation of r equals $(1 - \rho^2)/\sqrt{n}$. Based on experimental sampling Gosset ("Student" [255]) guessed that

$$p(r|\rho = 0) = (1 - r^2)^{(n-4)/2}/B(\frac{1}{2},\frac{1}{2}(n - 2)).$$

He also tried to find $p(r)$ for $\rho = 0.66$ but did not succeed.

H.E. Soper (1865–1930), working in Pearson's Department, used the δ-method to find the mean and the standard deviation of r to a second approximation, that is,

$$E(r) = \rho \left(1 - \frac{1 - \rho^2}{2(n - 1)} + \dots\right)$$

and

$$\sigma(r) = \frac{1 - \rho^2}{\sqrt{n - 1}} \left(1 + \frac{11\rho^2}{4(n - 1)} + \dots\right).$$

The distribution of the correlation coefficient was thus a burning issue in 1913. Surprisingly the problem was not solved by a member of the statistical establishment but by the 24-year-old school teacher R.A. Fisher who sent a manuscript to Pearson in September 1914, which was published in *Biometrika* [63] as " Frequency Distribution of the Values of the Correlation Coefficient in Samples from an Indefinitely Large Population."

Fisher refers to the two papers by Gosset and to Soper. He says that the problem may be solved by means of geometrical ideas, "The only difficulty lies in the expression of an element of volume in $2n$-dimensional space in terms of these derivatives," the "derivatives" being the five statistics $n\overline{x}$, $n\overline{y}$, ns_1^2, ns_2^2, nrs_1s_2 (our notation). Let the projection of the sample point $P_1 = (x_1, \ldots, x_n)$ on the equiangular line in the n-dimensional space be $M_1 = (\bar{x}, \ldots, \bar{x})$. Fisher shows that the square of the length of the vector $M_1 P_1$ equals $n s_1^2$ and that

$$dx_1 \cdots dx_n \propto s_1^{n-2} ds_1 d\bar{x}.$$

Analogous relations hold for $P_2 = (y_1, \ldots, y_n)$ and M_2. Fisher remarks that r is the cosine of the angle between $M_1 P_1$ and $M_2 P_2$ and continues: "Taking one of the projections as fixed at any point on the sphere of radius $s_2 \sqrt{n}$ the region for which r lies in the range dr, is a zone, on the other sphere in $n-1$ dimensions, of radius $s_1 \sqrt{n}\sqrt{1-r^2}$, and of width $s_1 \sqrt{n} dr/\sqrt{1-r^2}$, and therefore having a volume proportional to $s_1^{n-2}(1-r^2)^{(n-4)/2} dr$." Hence

$$dx_1 \cdots dx_n dy_1 \cdots dy_n \propto d\bar{x} d\bar{y} s_1^{n-2} ds_1 s_2^{n-2} ds_2 (1-r^2)^{(n-4)/2} dr.$$

Introducing this result and the five statistics in the joint probability element

$$\prod_{i=1}^{n} p(x_i, y_i) dx_i dy_i,$$

it is easy to see that (1) the distribution of $(\bar{x}, \bar{y})$ is bivariate normal with correlation coefficient ρ, (2) $(\bar{x}, \bar{y})$ are independent of the three other statistics, and (3) the distribution of these statistics is given by

$$p(s_1, s_2, r) \propto s_1^{n-2} s_2^{n-2} (1-r^2)^{(n-4)/2} \exp\left[-\frac{n}{2(1-\rho^2)} \left(\frac{s_1^2}{\sigma_1^2} - \frac{2\rho r s_1 s_2}{\sigma_1 \sigma_2} + \frac{s_2^2}{\sigma_2^2} \right) \right].$$

$$(18.1)$$

To find $p(r)$, Fisher makes the transformation

$$v = \frac{s_1 s_2}{\sigma_1 \sigma_2}, \quad 0 \le v < \infty, \quad \text{and} \quad e^z = \frac{s_1 \sigma_2}{s_2 \sigma_1}, \quad -\infty < z < \infty,$$

which leads to

$$p(v, z, r) \propto (1-r^2)^{(n-4)/2} v^{n-2} \exp\left(-\frac{nv}{1-\rho^2}(\cosh z - \rho r)\right),$$

whereafter integration with respect to v and z gives

$$p(r) \propto (1-r^2)^{(n-4)/2} \int_0^\infty (\cosh z - \rho r)^{-(n-1)} dz.$$

To evaluate the integral, Fisher sets $-\rho r = \cos\theta$, and differentiating the formula

$$\int_0^\infty (\cosh z + \cos\theta)^{-1} dz = \frac{\theta}{\sin\theta},$$

he gets

$$I_{n-1}(\rho r) = \int_0^\infty (\cosh z + \cos \theta)^{-(n-1)} dz$$

$$= \frac{1}{(n-2)!} \left(\frac{\partial}{\sin \theta \partial \theta}\right)^{n-2} \left(\frac{\theta}{\sin \theta}\right), \ \theta = \cos^{-1}(-\rho r)$$

Hence

$$p(r) = \frac{(1-\rho^2)^{(n-1)/2}}{\pi(n-3)!}(1 - r^2)^{(n-4)/2} \left(\frac{\partial}{\sin \theta \partial \theta}\right)^{n-2} \left(\frac{\theta}{\sin \theta}\right),$$

$$-1 \leq r \leq 1, \ n \geq 2$$

This remarkable result shows that $p(r)$ is a finite sum of trigonometric functions of θ. Fisher points out that the shape of the frequency curve depends strongly on n and ρ, and that even for high values of n the curve will be skew if $|\rho|$ is large. Therefore the values of the mean and the standard deviation of r cease to have any useful meaning, and

> It would appear essential in order to draw just conclusions from an observed high value of the correlation coefficient, say .99, that the frequency curves should be reasonably constant in form.

So far the paper is uncontroversial. It demonstrates Fisher's extraordinary mathematical powers, both geometrical and analytical. However, the formulation of the last two pages on the estimation of ρ turned out to be ambiguous. Fisher points out that the estimate obtained by setting $r = E(r)$ will be different from that following from $f(r) = E\{f(r)\}$, unless f is linear, and continues:

> I have given elsewhere [62] a criterion, independent of scaling, suitable for obtaining the relation between an observed correlation of a sample and the most probable value of the correlation of the whole population. Since the chance of any observation falling in the range of dr is proportional to

$$(1 - \rho^2)^{(n-1)/2}(1 - r^2)^{(n-4)/2} \left(\frac{\partial}{\sin \theta \partial \theta}\right)^{n-1} \frac{\theta^2}{2} dr$$

> for variations of ρ, we must find that value of ρ for which this quantity is a maximum.

Setting the derivative equal to zero and solving for ρ, he gets to a first approximation

$$r = \hat{\rho} \left\{1 + \frac{1 - r^2}{2n}\right\},$$

so that "the most likely value of the correlation will in general be less than that observed."

Fisher's paper naturally set Pearson at work. In "A Cooperative Study" by Soper, Young, Cave, Lee, and Pearson [241] they published 51 pages of mathematical investigations of $p(r)$ and 35 pages of tables, an enormous amount of work.

Section 8 of the paper is headed "On the Determination of the 'Most Likely' Value of the Correlation in the Sampled Population,"(i.e. $\hat{\rho}$.) They write:

> Fisher's equation, our (lxi), is deduced under the assumption that $\phi(\rho)$ is constant. In other words he assumes a horizontal frequency curve for ρ, or holds that *a priori* all values of ρ are equally likely to occur. This raises a number of philosophical points and difficulties.

They study the posterior distribution of ρ for various prior distributions and reach the well-known result that for large samples the choice of prior does not matter much, whereas the opposite is true for small samples. In particular, they give the first four terms of an expansion for $\hat{\rho}$ in terms of r and powers of $(n-1)^{-1}$ for a uniform prior, noting that the first two terms agree with Fisher's result. They argue against the uniform prior because experience shows that ordinarily ρ is not distributed in this way.

Fisher came to know of this misrepresentation of his method of estimation at a time when it was too late to correct it. It is easy to see how the misunderstanding could evolve because Fisher both in 1912 and 1915 used the terminology of inverse probability to describe the new method, which he later called the method of maximum likelihood. In 1920 he submitted an answer to the criticism to *Biometrika*, but Pearson replied that "I would prefer you published elsewhere."

Fisher's paper [66], published in *Metron*, contains a "Note on the Confusion Between Bayes's Rule and My Method of the Evaluation of the Optimum." For the first time Fisher explains unequivocally the distinction between the posterior mode and the maximum likelihood estimate. He writes:

> What we can find from a sample is the *likelihood* of any particular value of ρ, if we define the likelihood as a quantity proportional to the probability that, from a population having that particular value of ρ, a sample having the observed value r, should be obtained. So defined, probability and likelihood are quantities of an entirely different nature.

He concludes that "No transformation can alter the value of the optimum [the maximum likelihood] or in any way affect the likelihood of any suggested value of ρ." With hindsight we can see that these statements are implied by the formulations in his previous two papers.

The main content of the paper is a discussion of the transformation $r = \tan z$, or

$$z = \frac{1}{2} \ln \frac{1+r}{1-r} \text{ and } \zeta = \frac{1}{2} \ln \frac{1+\rho}{1-\rho}, \tag{18.2}$$

and the corresponding distribution

$$p(z) = \frac{n-2}{\pi} \operatorname{sech}^{n-1} \zeta \operatorname{sech}^{n-2} z I_{n-1}(\rho r), \quad -\infty < z < \infty.$$

Fisher derives a series expansion for $p(z)$ in terms of powers of $z - \zeta$ and $(1-n)^{-1}$, from which he obtains

$$E(z) = \zeta + \frac{\rho}{2(n-1)} + \cdots$$

and

$$V(z) = \frac{1}{n-1}\left(1 + \frac{4-\rho^2}{2(n-1)} + \cdots\right) = \frac{1}{n-3}\left(1 - \frac{\rho^2}{2(n-1)} - \cdots\right),$$

and similar series for μ_3 and μ_4, in all cases including terms of the order $(n-1)^{-2}$. It follows that

$$\beta_1 \cong \frac{\rho^2}{(n-1)^3}\left(\rho^2 - \frac{9}{16}\right)^2 \text{ and } \beta_2 \cong 3 + \frac{32 - 3\rho^4}{16(n-1)},$$

so that $p(z)$ tends rapidly to normality. Fisher notes that the value of ρ has very little influence on the shape of the distribution. He concludes:

> When expressed in terms of z, the curve of random sampling is therefore sufficiently normal and constant in deviation to be adequately represented by a probable error.

Hence the complicated distribution of r may miraculously in most cases be approximated by the normal distribution of z with mean $\zeta + \rho/2(n-1)$ and standard deviation $1/\sqrt{n-3}$. At a stroke Fisher had thus made the numerical results in the "cooperative study" practically superfluous.

Fisher remarks that for a single sample the correction $\rho/2(n-1)$ may usually be disregarded because it is small compared with the standard deviation. However, when averages of two or more samples are calculated, the correction should be included. He shows how the z transformation can be used for testing the significance of a value of r and the difference between two values of r using the traditional normal theory. A full discussion with examples is provided in *Statistical Methods* [73].

Fisher does not explain how he had derived the transformation, apart from stating that it leads to a nearly constant standard deviation. He has obviously used the formula $\sigma\{f(r)\} \cong \sigma(r)f'(r)$, which shows that for $\sigma(f)$ to be constant f' has to be proportional to $(1 - r^2)^{-1}$.

After these fundamental results on the distribution of the correlation coefficient, Fisher extended his analysis to the partial and multiple correlation coefficients. We indicate his main results. Let us denote the second-order sample moments by

$$s_{11} = s_1^2, \ s_{22} = s_2^2, \ s_{12} = rs_1s_2,$$

corresponding to the population values $\{\sigma_{ij}\}$ with the inverse

$$\sigma^{11} = \frac{1}{\sigma_1^2(1-\rho^2)}, \ \sigma^{22} = \frac{1}{\sigma_2^2(1-\rho^2)}, \ \sigma^{12} = \frac{-\rho}{\sigma_1\sigma_2(1-\rho^2)}.$$

Fisher [76] remarks that the distribution of the sample moments can be obtained from (18.1) by the above transformation, which gives

$$p(s_{11}, s_{22}, s_{12}) = \frac{n^{n-1}}{4\pi(n-3)!}(\sigma^{11}\sigma^{22} - \sigma_{12}^2)^{(n-1)/2}(s_{11}s_{22} - s_{12}^2)^{(n-4)/2}$$

$$\times \exp\left\{-\tfrac{1}{2}n(\sigma^{11}s_{11} + \sigma^{22}s_{22} + 2\sigma^{12}s_{12})\right\}, \tag{18.3}$$

$$s_{11} \ge 0, \ s_{22} \ge 0, \ s_{12}^2 \le s_{11}s_{22}.$$

The generalization to higher dimensions is due to Wishart [268].

Let (x_1, x_2, x_3) be normally correlated, and consider the conditional distribution $p(x_1, x_2 | x_3)$ which is bivariate normal. It is easy to see that the correlation between x_1 and x_2 for given x_3 is

$$\rho_{12.3} = \frac{\rho_{12} - \rho_{13}\rho_{23}}{\{(1-\rho_{13}^2)(1-\rho_{23}^2)\}^{1/2}}, \tag{18.4}$$

which is called the partial correlation coefficient. It does not depend on x_3.

Let (x_{i1}, x_{i2}, x_{i3}), $i = 1, \ldots, n$, be a sample of n independent observations. Eliminating the average effect of x_3 on x_1 by means of the regression, the residuals are

$$e_{i1.3} = x_{i1} - \overline{x}_1 - b_{13}(x_{i3} - \overline{x}_3), \ b_{13} = \frac{r_{13}s_1}{s_3},$$

a similar formula being valid for $e_{i2.3}$. We denote the vector of deviations $x_{i1} - \overline{x}_1$ and the vector of residuals $\{e_{i1.3}\}$ by $x_1 - \overline{x}_1$ and $e_{1.3}$, respectively. Using the Gaussian summation notation, we have $[e_{1.3}] = 0$,

$$s_{1.3}^2 = \frac{[e_{1.3}e_{1.3}]}{n} = s_1^2(1 - r_{13}^2)$$

and

$$s_{12.3} = \frac{[e_{1.3}e_{2.3}]}{n} = (r_{12} - r_{13}r_{23})s_1s_2,$$

so that the sample partial correlation coefficient becomes

$$r_{12.3} = \frac{[e_{1.3}e_{2.3}]}{\{[e_{1.3}e_{1.3}][e_{2.3}e_{2.3}]\}^{1/2}}, \tag{18.5}$$

which may be written in the same form as (18.4) with r instead of ρ. Moreover,

$$[e_{1.3}(x_3 - \overline{x}_3)] = [e_{2.3}(x_3 - \overline{x}_3)] = 0, \tag{18.6}$$

which is the well-known result that the residual is orthogonal to the independent variable.

Using the same representation of the sample in n-dimensional space as in 1915, Fisher [70] remarks that the lengths of the three vectors OP_1, OP_2, and OP_3 are proportional to the standard deviations of the three variables and that the correlation coefficients are the cosines of the angles between these vectors. It follows from (18.6) that the vectors $e_{1.3}$ and $e_{2.3}$ are orthogonal to OP_3 and from (18.5) that $r_{12.3}$ equals the cosine of the angle between $e_{1.2}$ and $e_{1.3}$. Hence, by projecting OP_1 and OP_2 on the $(n-1)$-dimensional space orthogonal to OP_3, we get two points Q_1 and Q_2, say, such that $r_{12.3}$ is the cosine of the angle Q_1OQ_2. The distribution of $r_{12.3}$ is thus the same as that of r_{12} with n replaced by $n-1$. Conditioning on more variables, the same argument can be used to prove that the resulting partial correlation coefficient is distributed as r_{12} for a sample size equal to n minus the number of conditioning variables.

The completion of Fisher's research on the sampling distributions of statistics under normality came with his paper [79] "The General Sampling Distribution of the Multiple Correlation Coefficient." He announces the importance of this paper as follows.

> Of the problem of the exact distribution of statistics in common use that of the multiple correlation coefficient is the last to have resisted solution. It will be seen that the solution introduces an extensive group of distributions, occurring naturally in the most diverse types of statistical investigations, and which in their mere mathematical structure supply an extension in a new direction of the entire system of distributions previously obtained, which in their generality underlie the analysis of variance.

Let $(y, x_1, \ldots, x_m)$ be normally correlated. The multiple correlation coefficient $\overline{R}$, $0 \le \overline{R} \le 1$, is defined by the relation

$$\sigma_{y.\underline{x}}^2 = \sigma_y^2(1 - \overline{R}^2), \tag{18.7}$$

where the left side denotes the variance in the conditional distribution of y for given values of the remaining m variables, denoted by $\underline{x}$. The sample multiple correlation coefficient R, $0 \le R \le 1$, is defined analogously to (18.7) by the equation

$$s_{y.\underline{x}}^2 = s_y^2(1 - R^2), \tag{18.8}$$

where $s_{y.\underline{x}}^2$ is the residual variance of y after correcting for the influence of $\underline{x}$ by the least squares regression. Hence

$$R^2 = \frac{\sum_{i=1}^n (y_i - \overline{y})(Y_i - \overline{y})}{\sum_{i=1}^n (y_i - \overline{y})^2}. \tag{18.9}$$

For $m = 1$ we have $R = |r|$.

Using his extraordinary geometrical insight Fisher finds

$$p(R^2) = \frac{(1 - \overline{R}^2)^{m_1 + m_2}}{\Gamma(m_1 + m_2)} (R^2)^{m_1 - 1} (1 - R^2)^{m_2 - 1} \sum_{j=0}^{\infty} \frac{\Gamma(m_1 + m_2 + j)}{B(m_1 + j, m_2)} \frac{(\overline{R}^2 R^2)^j}{j!},$$

(18.10)

where $m_1 = \frac{1}{2}m$, $m_2 = \frac{1}{2}(n - m - 1)$, and $m_1 + m_2 = \frac{1}{2}(n - 1)$.

To find the asymptotic distribution of R^2 for m fixed and $n - m - 1$ tending to infinity, Fisher sets

$$B^2 = (n - m - 1)R^2 \text{ and } \beta^2 = (n - m - 1)\overline{R}^2.$$

Noting that

$$(1 - R^2)^{(n - m - 1)/2} \to e^{-B^2/2},$$

and using Stirling's formula, (18.10) gives

$$p(B^2) = e^{-\beta^2/2 - B^2/2} \frac{(B^2)^{m_1 - 1}}{2^{m_1}} \sum_{j=0}^{\infty} \frac{(\beta^2 B^2)^j}{\Gamma(m_1 + j)2^{2j} j!}.$$

(18.11)

Today the distribution of B^2 is called the noncentral χ^2 distribution with m degrees of freedom and noncentrality parameter β^2.

18.4 The Sufficiency of the Sample Variance, 1920

Assuming normality, Fisher [65] derives the distribution of the sample variance and the first two moments of the distribution of the mean deviation. He did not know that these results are due to Helmert [122]; see Section 17.2. He discusses the relative efficiency of the estimates of σ obtained from the absolute moments of any order and proves that the estimate based on the second moment is most efficient; he did not know that this result is due to Gauss [102]. Let s_1 and s_2 denote the estimates of σ based on the mean deviation and the standard deviation, respectively. Like Laplace [162] (see Section 12.3), Fisher investigates the joint and conditional distributions to find out which estimate is the better. Because $p(s_1)$ is complicated, he considers only the case $n = 4$, and even in this case the distribution depends critically on the configuration of the sample. However, Fisher reaches the following results.

> From the manner in which the frequency surface has been derived, as in expressions III, it is evident that: *For a given value of s_2, the distribution of s_1 is independent of σ.* On the other hand, it is clear from expressions (IV.) and (V.) that for a given value of s_1 the distribution of s_2 does involve σ. In other words, if, in seeking information as to the value of σ, we first determine s_1, then we can still further improve our estimate by determining s_2; but if we had first determined s_2,

the frequency curve of s_1 being entirely independent of σ, the actual value of s_1 can give us no further information as to the value of σ. The whole of the information to be obtained from s_1 is included in that supplied by a knowledge of s_2.

This remarkable property of s_2, as the methods that we have used to determine the frequency surface demonstrate, follows from the distribution of frequency density in concentric spheres over each of which s_2 is constant. It therefore holds equally if s_3 or any other derivate be substituted for s_1. If this is so, then it must be admitted that:

The whole of the information respecting σ, which a sample provides, is summed up in the value of s_2.

[We have changed Fisher's notation from σ_1 and σ_2 to s_1 and s_2.]

Two years later Fisher [67] introduced the term "sufficiency" for "this remarkable property of s_2."

Finally, Fisher remarks that this property depends on the assumption of normality. If instead the observations are distributed according to the double exponential

$$\frac{1}{\sigma\sqrt{2}} \exp\left(-|x - m|\sqrt{2}/\sigma\right),$$

then "s_1 may be taken as the ideal measure of σ."

The Revolutionary Paper, 1922

19.1 The Parametric Model and Criteria of Estimation, 1922

During the period 1912–1921 Fisher had, at least for himself, developed new concepts of fundamental importance for the theory of estimation. He had rejected inverse probability as arbitrary and leading to noninvariant estimates; instead he grounded his own theory firmly on the frequency interpretation of probability. He had proposed to use invariance and the method of maximum likelihood estimation as basic concepts and had introduced the concept of sufficiency by an important example. Thus prepared, he was ready to publish a general theory of estimation, which he did in the paper [67], "On the Mathematical Foundations of Theoretical Statistics." For the first time in the history of statistics a framework for a frequency-based general theory of parametric statistical inference was clearly formulated.

Fisher says that the object of statistical methods is the reduction of data, which is accomplished by considering the data at hand as a random sample from a hypothetical infinite population, whose distribution with respect to the characteristics under discussion is specified by relatively few parameters. He divides the problems into the following types which he formulates as follows ([67], p. 313).

(1) Problems of Specification. These arise in the choice of the mathematical form of the population.

(2) Problems of Estimation. These involve the choice of methods of calculating from a sample statistical derivates, or as we call them statistics, which are designed to estimate the values of the parameters of the hypothetical population.

(3) Problems of Distribution. These include discussions of the distribution of statistics derived from samples, or in general any functions of quantities whose distribution is known.

The remaining part of the paper is taken up with a discussion of the problems of estimation and distribution. He defines three criteria of estimation: consistency, efficiency, and sufficiency ([67], pp. 309–310).

Consistency: A statistic satisfies the criterion of consistency, if, when it is calculated from the whole population, it is equal to the required parameter.

Efficiency: The efficiency of a statistic is the ratio (usually expressed as a percentage) which its intrinsic accuracy bears to that of the most efficient statistic possible. It expresses the proportion of the total available relevant information of which that statistic makes use.

Efficiency (Criterion): The criterion of efficiency is satisfied by those statistics that, when derived from large samples, tend to a normal distribution with the least possible standard deviation.

Sufficiency: A statistic satisfies the criterion of sufficiency when no other statistic that can be calculated from the same sample provides any additional information as to the value of the parameter to be estimated.

Fisher's precise definition of these concepts is a breakthrough in the theory of estimation; from then on it became a standard practice to discuss the properties of estimates in relation to the three criteria.

In the following discussion of large-sample estimation theory, Fisher considers the class of statistics that are asymptotically normal with mean θ and variance σ^2/n . He thus follows the tradition grounded in Laplace's central limit theorem. However, he also indicates a possible extension of this theory by considering the "intrinsic accuracy" and the "relevant information" of a statistic.

Fisher ([67], p. 317) begins by using the criterion of sufficiency. Let t_1 be sufficient and t_2 any other statistic so that the factorization

$$p(t_1, t_2 | \theta) = p(t_1 | \theta)p(t_2 | t_1), \tag{19.1}$$

where $p(t_2 | t_1)$ does not depend on θ, holds for any t_2 different from t_1. He assumes that (t_1, t_2) are asymptotically normal $(\theta, \theta, \sigma_1^2, \sigma_2^2, \rho)$, with variances of order $1/n$. From the conditional distribution it follows that

$$E(\, t_2 | \, t_1) = \theta + \frac{\rho \sigma_2}{\sigma_1}(t_1 - \theta)$$

and $V(t_2 | t_1) = \sigma_2^2(1 - \rho^2)$. The condition for the conditional distribution to be independent of θ is thus that $\rho \sigma_2 = \sigma_1$, which implies that σ_1 is less than σ_2 and that the efficiency of t_2 is ρ^2. Hence an asymptotically normal sufficient estimate has minimum variance within the class of estimates considered. It is shown that Fisher uses the same method of proof for finding the relative efficiency as Laplace.

Fisher adds:

> Besides this case we shall see that the criterion of sufficiency is also applicable to finite samples, and to those cases when the weight of a statistics is not proportional to the number of the sample from which it is calculated.

The three types of problems and the three criteria of estimation give the framework for a research program that came to dominate theoretical statistics for the rest of the century. Before continuing with Fisher's mathematical contributions we discuss another important aspect of his work, namely the creation of a new technical vocabulary for mathematical statistics. We have used this terminology throughout the previous chapters to explain the older concepts in a way easily understandable to modern readers.

In connection with the specification, he introduces the term "parameter", so that today we speak of a parametric family of distributions and parametric statistical inference. It is characteristic for his style of writing that he usually does not include such (obvious) mathematical details as the definition of the parameter space and the sample space.

He coined the word "statistic", which naturally caused much opposition, for a function of the sample, designed to estimate the value of a parameter or to test the goodness of fit. As a natural consequence he speaks of the sampling distribution of a statistic. When he needed an expression for the square of Pearson's "standard deviation", he did not use Edgeworth's "fluctuation" but introduced the term "variance" [64] from which flowed "the analysis of variance."

We have already discussed his definition of the term "likelihood" followed by the method of maximum likelihood. He also used the name "ideal score" for the derivative of the log-likelihood function. To supplement the maximum likelihood estimate, he introduced "ancillary statistics."

He characterized estimates as discussed above by the properties of consistency, efficiency, and sufficiency, and introduced the concept of "information" in the sample and in an estimate. He coined the terms null hypothesis, test of significance, level of significance, and percentage point. In the design of experiments he introduced randomization, factorial designs, and interaction as new concepts.

It is clear that today we cannot discuss statistical theory without making use of the Fisherian terminology. David [40] has collected a list of the "First (?) Occurrence of Common Terms in Mathematical Statistics."

A related matter of great importance is Fisher's sharp distinction between sample and population values, both verbally and notationally. This distinction occurs of course also in works by previous authors, who write on the mean, say, and its true value, but the term sample versus population value is convenient and unequivocal. Moreover, in contradistinction to the then-prevailing practice, Fisher gradually came to use different symbols for the two concepts, which eventually resulted in the use of Latin letters for sample values and Greek letters for population values.

19.2 Properties of the Maximum Likelihood Estimate

As discussed in Section 11.6, Fisher uses the binomial distribution parameterized in the usual way and alternatively by the transformation $\sin \zeta = 2\theta - 1$ to demonstrate that the principle of inverse probability leads to different estimates of θ depending on which function of θ is considered uniformly distributed. He was not the first to do so, but he was the first to propose an alternative method of estimation, the method of maximum likelihood, that is invariant to parameter transformations and has the same asymptotic properties as the method he rejected.

Fisher's proof of the properties of the maximum likelihood estimate is as follows ([67], pp. 328–330). To find the distribution of $\hat{\theta}$ in terms of $f(x|\theta)$, he uses the fact that

$$p(\hat{\theta}|\theta) = \int_R p(\underline{x}|\theta)d\underline{x} \text{ for } R = \underline{x} : \hat{\theta}(\underline{x}) = \theta, \tag{19.2}$$

where

$$\ln p(\underline{x}|\theta) = \sum \ln f(x_i|\theta).$$

Expanding $\ln f(x|\theta)$ in Taylor's series about $\theta = \hat{\theta}$ and using the fact that

$$l'(\theta) = \sum \frac{\partial}{\partial \theta} \ln f(x_i|\theta) = 0 \text{ for } \theta = \hat{\theta},$$

he finds

$$\ln p(\underline{x}|\theta) = \ln p(\underline{x}|\hat{\theta}) + \frac{1}{2}(\theta - \hat{\theta})^2 \sum \frac{\partial^2 \ln f(x_i|\hat{\theta})}{\partial \theta^2} + \cdots \tag{19.3}$$

$$= \ln p(\underline{x}|\hat{\theta}) + \frac{1}{2}n\bar{b}(\theta - \hat{\theta})^2 + \cdots,$$

where

$$b_i = \frac{\partial^2 \ln f(x_i|\hat{\theta})}{\partial \theta^2}.$$

Assuming that $\hat{\theta} - \theta$ is of order $n^{-1/2}$ and noting that for sufficiently large samples $\sum b_i$ differs from $n\bar{b}$ only by a quantity of order $\sigma_b\sqrt{n}$, he concludes by inserting (19.3) in (19.2) that $\hat{\theta}$ is asymptotically normal $\{\theta, 1/(-n\bar{b})\}$. He writes (p. 329): "The formula

$$-\frac{1}{\sigma_\theta^2} = n\overline{\frac{\partial^2}{\partial \theta^2} \log f} \tag{19.4}$$

supplies the most direct way known to me of finding the probable errors of statistics. It may be seen that the above proof applies only to statistics obtained by the method of maximum likelihood."

The same formula had been derived by Laplace [150] by inverse probability; see Section 5.6.

In applications of (19.4) Fisher replaces the average by its expectation. In our notation $-n\bar{b} = j(\hat{\theta})$. By the substitution $\hat{\theta} = \theta + O(n^{-1/2})$ Fisher had presumably found that $j(\hat{\theta}) = j(\theta) + O(n^{-1/2})$, whereafter an application of the central limit theorem leads to the approximation $j(\hat{\theta}) = i(\theta) + O(n^{-1/2})$. Anyway, in his examples he sets $V(\hat{\theta}) = 1/i(\theta)$. He remarks that analogous formulas hold for several parameters and displays the formulas for two parameters.

The main tool in his proof is the expansion (19.3) which previously had been used by Edgeworth. However, Fisher is entirely within the tradition of direct probability by using (19.3) to find the sampling distribution of $\hat{\theta}$, whereas Edgeworth derives the posterior distribution of θ. Nevertheless, it is strange that Fisher does not refer to Edgeworth; he seems to have overlooked Edgeworth's paper not only in 1912 but also in 1922.

Fisher ([67], p. 330) gives an unsatisfactory proof of the property that the maximum likelihood estimate has minimum variance within the class of asymptotically normal estimates with expectation θ. We here relate the improved version of his proof ([74], p. 711).

For $l(\theta) = \ln\{p(\underline{x}|\theta)/p(\underline{x}|\hat{\theta})\}$ we have from (19.3) that

$$l'(\theta) = -i(\theta)(\theta - \hat{\theta}) + \cdots .$$

From the asymptotic normality of t it follows that

$$\sigma_t^{-2} = -\frac{\partial^2 \ln p(t|\theta)}{\partial \theta^2} = \left(\frac{p'}{p}\right)^2 - \frac{p''}{p} + \cdots .$$

Writing $p(t|\theta)$ in the form

$$p(t|\theta) = \int_R p(\underline{x}|\theta)d\underline{x} \text{ for } R = \{\underline{x} : t(\underline{x}) = t\},$$

and using the approximation above, Fisher finds

$$\frac{p'}{p} = \frac{\int_R [\partial \ln p(\underline{x}|\theta)/\partial \theta] p(\underline{x}|\theta)d\underline{x}}{\int_R p(\underline{x}|\theta)d\underline{x}}$$

$$= \frac{i(\theta) \int_R (\hat{\theta} - \theta) p(\underline{x}|\theta)d\underline{x}}{\int_R p(\underline{x}|\theta)d\underline{x}}$$

$$= i(\theta) E(\hat{\theta} - \theta|t).$$

Using the fact that

$$\frac{p''}{p} = \left(\frac{p'}{p}\right)^2 + \frac{\partial^2 \ln p}{\partial \theta^2},$$

a similar reasoning gives

$$\frac{p''}{p} = i^2(\theta) E\{(\hat{\theta} - \theta)^2|t\} - i(\theta),$$

which leads to Fisher's formula

$$\sigma_t^{-2} = i(\theta) - i^2(\theta) V(\hat{\theta}|t).$$

Because $V(\hat{\theta}|t) = 0$ for $t = \hat{\theta}$ and positive otherwise it follows that σ_t^2 is minimized and takes on the value $1/i(\theta)$ for $t = \hat{\theta}$.

One of Fisher's favorite examples is the Cauchy distribution. Like Poisson he points out that the distribution of the arithmetic mean is the same as for the individual observations. Like Laplace he notes that the variance of the median is $\pi^2/4n$ if the scale parameter equals unity. He does not mention Poisson and Laplace. He derives the maximum likelihood equation and $V(\hat{\theta}) = 2/n$, which shows that the efficiency of the median is $8/\pi^2$.

For Pearson's Type III distribution with three parameters he proves, like Edgeworth, that the efficiency of the estimate of the location parameter, based on the arithmetic mean, equals $(p-1)/(p+1)$, $p > 1$. However, he continues the analysis by deriving the maximum likelihood equations for all three parameters and the corresponding dispersion matrix, which shows that the method of moments leads to inefficient estimates.

Edgeworth had derived the maximum likelihood equations and the variance of the estimates for the location family of distributions and for the scale family separately, and noted that a similar analysis could be carried out for the location-scale family. Fisher ([67], pp. 338–342) carries out this analysis.

Turning to discrete distributions, he derived the maximum likelihood estimates and their variance for the parameters in the binomial and Poisson distributions. For the multinomial distribution he ([67], pp. 357–358) set

$$-l(\theta) = \sum_{i=1}^{k} x_i \ln\left(\frac{x_i}{m_i}\right), \quad \sum x_i = \sum m_i = n, \quad m_i = np_i(\theta).$$

Assuming that n is large and that $e_i = x_i - m_i$ is small compared with m_i for all i, he finds

$$-l(\hat{\theta}) = \sum (m_i + e_i) \ln \frac{m_i + e_i}{m_i} = \frac{1}{2} \sum \frac{e_i^2}{m_i} + \cdots .$$

Hence, under the conditions stated, $-2l(\hat{\theta}) = \chi^2$ to a first approximation, which has the advantage that the distribution of $l(\hat{\theta})$ then is known and tabulated. If the approximation breaks down, $l(\hat{\theta})$ should be used as a test for goodness of fit, and it is therefore desirable to study its distribution. It should be noted that the equation for determining the optimum value of θ by minimizing χ^2 is

$$\sum \frac{x_i^2 m_i'}{m_i^2} = 0,$$

whereas the likelihood equation is

$$\sum \frac{x_i m_i'}{m_i} = 0.$$

Fisher ([67], pp. 359–363) uses the method of maximum likelihood to find the variance of the mean and standard deviation calculated from grouped normal data, when the group interval is of length $a\sigma$. To a first approximation his results are

$$V(\hat{\mu}) = \frac{\sigma^2}{n}\left(1 + \frac{a^2}{12}\right) \text{ and } V(\hat{\sigma}) = \frac{\sigma^2}{2n}\left(1 + \frac{a^2}{6}\right),$$

from which the efficiencies are easily calculated. For $a = \frac{1}{2}$, say, the efficiency of the grouped mean is about 0.98, and for the grouped standard deviation, about 0.96.

However, as remarked by Fisher (p. 363):

> Although for the normal curve the loss of efficiency due to moderate grouping is very small, such is not the case with curves making a finite angle with the axis, or having at an extreme a finite or infinitely great ordinate. In such cases even moderate grouping may result in throwing away the greater part of the information which the samples provides.

In his discussion of the Pearsonian distributions, Fisher (pp. 348–351) points out that for frequency curves having a finite angle with the axis it is more accurate for large samples to locate the curve by the extreme observations than by the mean. He illustrates this by means of the uniform distribution over the interval $\mu \pm \frac{1}{2}a$. It is easy to find the distribution of the smallest and the largest observation, and noting that for large n these statistics are independent, Fisher derives the distribution of the midrange $t = \frac{1}{2}(x_{(1)} + x_{(n)})$,

$$p(t) = \frac{n}{\alpha} \exp\left\{ -\frac{2n\,|t - \mu|}{\alpha} \right\},$$

from which follows that $E(t) = \mu$ and $V(t) = \alpha^2/2n^2$; that is, the variance is of order n^{-2} instead of "as usual" of order n^{-1}. The distribution of the mean is asymptotically normal $(\mu, \alpha^2/12n)$. Fisher remarks:

> The two error curves are thus of radically different form, and strictly no value for the efficiency can be calculated; if, however, we consider the ratio of the two standard deviations, then

$$\frac{\sigma_{\hat{\mu}}^2}{\sigma_{\bar{x}}^2} = \frac{\alpha^2/2n^2}{\alpha^2/12n} = \frac{6}{n},$$

> when n is large, a quantity which diminishes indefinitely as the sample is increased [our notation].

This leads him to the further remark that it is desirable to develop a theory of efficiency valid for statistics having different limiting distributions and valid also for finite samples. This is the main topic of the paper [74]. However, some problems were only partially solved or just mentioned in [74], and he therefore returned to them in the papers "Two New Properties of Mathematical Likelihood" [82] and "The Logic of Inductive Inference" [84]. During the period 1922–1935 he moved from the method of maximum likelihood to a discussion of the complete likelihood function.

19.3 The Two-Stage Maximum Likelihood Method and Unbiasedness

Fisher does not use the criterion of unbiasedness because it leads to noninvariant estimates. Of course, he uses unbiased estimates but for different reasons. For the linear normal model he partitions the sum of squared errors into independent parts with a corresponding partition of the number of observations, as for example,

$$[\varepsilon\varepsilon] = [ee] + (b - \beta)'X'X(b - \beta) \text{ and } n = (n - m) + m,$$

from which three unbiased estimates of σ^2 are obtained.

The maximum likelihood estimate of the variance of the normal distribution is $\hat{\sigma}^2 = \sum (x_i - \overline{x})^2/n$ with the expectation $\sigma^2 - \sigma^2/n$. To obtain an unbiased estimate he uses the two-stage method of maximum likelihood ([63], [66]). For the distribution of $\hat{\sigma}^2$ we have

$$p(\hat{\sigma}^2|\sigma^2) \propto (\sigma^2)^{-(n-1)/2}\exp(-n\hat{\sigma}^2/2\sigma^2).$$

Maximizing this expression with respect to σ^2 we get

$$\hat{\sigma}^2 = \hat{\sigma}^2 n/(n - 1) = s^2,$$

which is unbiased for σ^2.

Considering k samples of size n from populations with different means but the same variance, the maximum likelihood estimates are the arithmetic means and the variance

$$\hat{\sigma}^2 = \sum_{i=1}^{k}\sum_{j=1}^{n}(x_{ij} - \overline{x}_i)^2/kn = (n - 1)s^2/n,$$

where

$$s^2 = \frac{1}{k}\sum_{i=1}^{k}s_i^2, \quad s_i^2 = \sum_{j=1}^{n}(x_{ij} - \overline{x}_i)^2/(n - 1).$$

For a fixed value of n and $k \to \infty$, s^2 tends in probability to σ^2 so that $\hat{\sigma}^2$ tends to $\sigma^2(n-1)/n$, which of course is unsatisfactory. This inconsistency of the maximum likelihood estimate was pointed out by Neyman and Scott [190]. They have, however, overlooked that Fisher ([63], [66]) in applying this method to normally distributed observations uses the two-stage method: First he maximizes the probability density of the observations and next he finds the sampling distribution of the estimates and for each estimate he maximizes its density, which leads to the final (unbiased) estimate. For the bivariate normal the second stage changes the maximum likelihood estimates of the variances by the factor $n/(n-1)$ as above, and it changes r to $\hat{\rho}$. Fisher does not mention the two-stage procedure in his paper [67].

Studentization, the F Distribution, and the Analysis of Variance, 1922–1925

20.1 Studentization and Applications of the t Distribution

Fisher never tired of emphasizing the importance of Gosset's idea; likewise he often repeated his own derivation of $p(\overline{x}, s^2)$. It was not until 1935 that he acknowledged Helmert's [120] priority with respect to the distribution of $[\varepsilon\varepsilon] = \sigma^2\chi^2$, he never mentioned that Helmert also derived $p(\overline{x}, s^2)$.

The importance and generality of the t distribution is explained by Fisher as follows: The t test involves two components both distributed independently of σ, namely $u = (\overline{x} - \mu)\sqrt{n}/\sigma$, which is normal $(0,1)$, and $\chi^2 = (n-1)s^2/\sigma^2$, which is distributed as χ^2 with $f = n - 1$, $(n-1)s^2 = \sum(x_i - \overline{x})^2$. It follows that

$$t = (\overline{x} - \mu)\frac{\sqrt{n}}{s} = \frac{u}{\sqrt{\chi^2/f}}. \tag{20.1}$$

Fisher [75] remarks that this formula

> ...shows that "Student's" formula for the distribution of t is applicable to all cases which can be reduced to a comparison of the deviation of a normal variate, with an independently distributed estimate of its standard deviation, derived from the sums of squares of homogeneous normal deviations, either from the true mean of the distribution, or from the means of samples.

This procedure later became known as "Studentization".

Fisher [75] demonstrates a general method for obtaining t-distributed statistics. He considers a random vector u of n independent normal $(0,1)$ variables and makes an orthogonal transformation to $v = Q'u$ so that $\sum u_i^2 = \sum v_i^2 = \chi^2$, $f = n$. Suppose that the first $m < n$ orthogonal vectors $q_1, \ldots, q_m$ are given; we can then always supplement these by a set of $n - m$ orthogonal vectors to obtain a complete set. It follows that the remainder

$$\sum_1^n u_i^2 - \sum_1^m v_i^2$$

can always be written as

$$\sum_{m+1}^n v_i^2,$$

which is distributed as χ^2, $f = n - m$, independently of $v_1, \ldots, v_m$. Hence we need only to check that the m given functions are orthonormal transformations of the us to be sure that the remainder is distributed as χ^2 with $f = n - m$. Setting

$$s^2 = \frac{1}{n - m} \sum_{m+1}^n v_i^2$$

we have that $t_i = v_i/s$, $i = 1, \ldots, m$, according to (20.1) are distributed as t with $f = n - m$.

As an example Fisher considers the linear model

$$y_i = \alpha + \beta(x_i - \overline{x}) + \sigma u_i, \ i = 1, \ldots, n,$$

for given values of $x_1, \ldots, x_n$. Estimating α and β by the method of least squares, we get $a = \overline{y}$,

$$b = \frac{\sum y_i(x_i - \overline{x})}{\sum (x_i - \overline{x})^2},$$

and

$$\sigma^{-2} \sum (y_i - \alpha - \beta(x_i - \overline{x}))^2 - \sigma^{-2} n(a - \alpha)^2 - \sigma^{-2} \sum (x_i - \overline{x})^2 (b - \beta)^2 \tag{20.2}$$

$$= \sigma^{-2} \sum (y_i - a - b(x_i - \overline{x}))^2.$$

It follows that the elements of q_1 all equal $n^{-1/2}$ and that the ith element of q_2 equals $(x_i - \overline{x})/\sum (x_i - \overline{x})^2$, and because q_1 and q_2 are orthonormal, the right side of (20.2) will be distributed as χ^2 with $f = n - 2$. Hence

$$s^2 = \frac{\sum (y_i - a - b(x_i - \overline{x}))^2}{n - 2}$$

is distributed as $\sigma^2 \chi^2/f$ independently of u and b, so the specified values of α and β can be tested by means of $t = (a - \alpha)\sqrt{n}/s$ and

$$t = (b - \beta) \frac{\sqrt{\sum (x_i - \overline{x})^2}}{s}, \tag{20.3}$$

respectively.

Fisher continues with the linear model with orthogonal components, the most important case being an expansion in terms of orthogonal polynomials, for which the above theory is immediately applicable. Finally he remarks that for the general linear model we have $V(b_r) = \sigma^2 k_{rr}$, where k_{rr} is the rth diagonal element of $(X'X)^{-1}$, and consequently

$$t_r = \frac{b_r - \beta_r}{s\sqrt{k_{rr}}}, \quad s^2 = \frac{\sum (y_i - \hat{\eta}_i)^2}{n - m}, \tag{20.4}$$

is distributed as t with $f = n - m$. Fisher's proof is incomplete because he does not orthogonalize the model to show that $s^2 = \sigma^2 \chi^2/(n - m)$ is independent of b_r. The orthogonalization may be found in Thiele [260].

20.2 The F Distribution

After having shown that the t-test may be used to test the significance of the difference between two means, Fisher turned to the corresponding problem for two variances. The usual practice so far had been to compare the difference $s_1^2 - s_2^2$ with its estimated standard deviation, a method that obviously is unsatisfactory for small samples. In a paper presented to the International Congress of Mathematicians in 1924, Fisher [78] points out this deficiency and proposes instead to use the variance ratio s_1^2/s_2^2, which under the null hypothesis is independent of the unknown σ^2.

Moreover, a discussion of the variance ratio test and its many applications were given in the last two chapters of *Statistical Methods* [73]. Fisher used the notation $e^{2z} = s_1^2/s_2^2$ and provided a table of percentage points of the z distribution for $P = 5\%$, wherefore the test became known as the z test until Snedecor [239] proposed to use $F = s_1^2/s_2^2$.

Writing the two independent sample variances as

$$s_1^2 = \frac{\sigma_1^2 \chi_1^2}{f_1} \text{ and } s_2^2 = \frac{\sigma_2^2 \chi_2^2}{f_2},$$

Fisher ([71], [78]) gets

$$e^{2z} = \frac{s_1^2/\sigma_1^2}{s_2^2/\sigma_2^2} = \frac{\chi_1^2/f_1}{\chi_2^2/f_2}.$$

Hence, the distribution of $F = e^{2z}$ is independent of σ_1^2 and σ_2^2. By means of the two χ^2 distributions, it is easy to prove that

$$p(F) = \frac{f_1^{f_1/2} f_2^{f_2/2}}{B\left(\frac{1}{2}f_1, \frac{1}{2}f_2\right)} \frac{F^{(f_1/2)-1}}{(f_1 F + f_2)^{(f_1+f_2)/2}}, \quad 0 \leq F < \infty,$$

from which we get

$$p(z) = \frac{2f_1^{f_1/2} f_2^{f_2/2}}{B\left(\frac{1}{2}f_1, \frac{1}{2}f_2\right)} \frac{e^{f_1 z}}{\left(f_1 e^{2z} + f_2\right)^{(f_1+f_2)/2}}, \quad -\infty < z < \infty,$$

which is the formula stated by Fisher.

Fisher's paper is entitled "A Distribution Yielding the Error Functions of Several Known Statistics." Accordingly he considers three limiting cases. For $f_2 \to \infty$, we have $\chi^2/f_2 \to 1$ so that $F \to \chi_1^2/f_1$. For $f_1 = 1$ we have $\chi_1^2/f_1 = u^2$, the square of a normal (0,1) variate, so that $F = t^2$ for $f = f_2$. For $f_1 = 1$ and $f_2 \to \infty$, F consequently becomes normally distributed. A survey of the derivation of these distributions and their relations to partial sums of the binomial and Poisson distributions can be found in Fisher [83].

The z or F distribution can obviously be used for testing whether an observed value of s_1^2/s_2^2 deviates significantly from a hypothetical value of σ_1^2/σ_2^2. Fisher points out that z is asymptotically normal with

$$V(z) = \frac{1}{2}\left(\frac{1}{f_1} + \frac{1}{f_2}\right),$$

because the variance of $\ln(\chi^2/f)$ equals $2/f$ for large values of f.

20.3 The Analysis of Variance

Fisher remarks that the z distribution, as do the t and χ^2 distributions, has many other applications than the simple one following directly from the definition of the statistics. He writes [78]:

> The practical working of cases involving the z distribution can usually be shown most simply in the form of an analysis of variance. If x is any value, $\overline{x}_p$ the mean of any class, and $\overline{x}$ the general mean, n the number of classes of s observations each, the following table shows the form of such an analysis:

Table 20.1. Analysis of variance for one way classification.

Variance	Degr. Freedom	Sums of Squares	Mean Square
Between classes	$n_1 = n - 1$	$sS_1^n(\overline{x}_p - \overline{x})^2$	s_1^2
Within classes	$n_2 = n(s - 1)$	$S_1^{ns}(x - \overline{x}_p)^2$	s_2^2
Total	$ns - 1$	$S_1^{ns}(x - \overline{x})^2$	

The two columns headed Degrees of Freedom and Sum of Squares must add up to the totals shown; the mean squares are obtained by dividing the sums of squares by the corresponding degrees of freedom
. . . .

This simple tabular representation of an analysis of variance is a pedagogical masterpiece that immediately found wide acceptance.

The second example is a test for the multiple correlation coefficient. Fisher writes the linear model in the form

$$E(y_i|\underline{x}) = \alpha + \sum_{j=1}^{m} \beta_j (x_{ij} - \overline{x}_j), \ i = 1, \ldots, n, \ V(y_i|\underline{x}) = \sigma^2 (1 - \overline{R}^2),$$

where $\overline{R}$ denotes the multiple correlation coefficient (our notation). The least squares estimate is

$$Y_i = \overline{y} + \sum_{j=1}^{m} b_j (x_{ij} - \overline{x}_j),$$

and the empirical multiple correlation coefficient R is defined by the relation

$$\frac{1}{n} \sum (y_i - Y_i)^2 = \frac{1}{n} \sum (y_i - \overline{y})^2 (1 - R^2).$$

Fisher presents the analysis of variance as in Table 20.2:

Table 20.2. Analysis of variance for regression.

Variance	Degr. Freedom	Sums of Squares
Of regression formula	m	$\sum (Y_i - \overline{y})^2 = ns^2 R^2$
Around regression	$n - m - 1$	$\sum (y_i - Y_i)^2 = ns^2 (1 - R^2)$
Total	$n - 1$	$\sum (y_i - \overline{y})^2 = ns^2$

Hence, the relation between F and R^2 becomes

$$F = \frac{\sum (Y_i - \overline{y})^2 / m}{\sum (y_i - Y_i)^2 / (n - m - 1)} = \frac{R^2 / m}{(1 - R^2)/(n - m - 1)},$$

or

$$R^2 = \frac{mF}{mF + n - m - 1},$$

so

$$p(R^2) = \frac{1}{B\left(\frac{1}{2}m, \frac{1}{2}(n - m - 1)\right)} (R^2)^{(m-2)/2} (1 - R^2)^{(n-m-3)/2}, \ 0 \le R^2 \le 1,$$

which is the distribution of R^2 under the assumption that $\overline{R}^2 = 0$. Another proof and discussion of this distribution can be found in Fisher [72].

The third example is a test for the significance of Pearson's [206] correlation ratio η. Suppose that for each value of the independent variable

x_i, $i = 1, \ldots, k$, we have n_i observations of the dependent variable with mean $\overline{y}_i$. Then

$$\frac{\eta^2}{1 - \eta^2} = \frac{\sum_{i=1}^{k} n_i(\overline{y}_i - \overline{y})^2}{\sum_{i=1}^{k} \sum_{j=1}^{n_i} (y_{ij} - \overline{y}_i)^2},$$

which is the ratio of two independent sums of squares, together giving the total sum of squares. From an analysis of variance analogous to the one for R^2, Fisher finds the distribution of η^2.

The fourth example is a test for the goodness of fit of a regression formula. Let the least squares estimate of the regression equation be

$$Y_i = \overline{y} + \sum_{j=1}^{m} b_j(x_{ij} - \overline{x}_j), \ i = 1, \ldots, k,$$

and let there be n_i observations of y for each i as above. Writing

$$y_{ij} - \overline{y} = (y_{ij} - \overline{y}_i) + (\overline{y}_i - Y_i) + (Y_i - \overline{y}),$$

squaring, and summing, the total sum of squares is partitioned into three components. Comparing the first two, Fisher gets

$$F = \frac{\sum_{i=1}^{k} n_i(\overline{y}_i - Y_i)^2/(k - m)}{\sum_{i=1}^{k} \sum_{j=1}^{n_i} (y_{ij} - \overline{y}_i)^2/(n - k)}, \ n = \sum_{i=1}^{k} n_i,$$

with $k - m$ and $n - k$ degrees of freedom. A significant value of F means that the proposed regression equation does not represent the data satisfactorily.

This epoch-making paper takes up only nine pages and was not printed [78] until four years after its presentation with the following Addendum by Fisher.

> Since the International Mathematical Congress (Toronto, 1924) the practical applications of the developments summarized in this paper have been more fully illustrated in the author's book *Statistical Methods for Research Workers* (Oliver and Boyd, Edinburgh, 1925). The Toronto paper supplies in outline the mathematical framework around which the book has been built, for a formal statement of which some reviewers would seem to have appreciated the need.

Statistical Methods [73] contains, except for some mathematical details, the results given above supplemented by many practical examples. There Fisher also shows how the same method can be used for partitioning the total variation of multiple classified data into independent components. In § 42 on "Analysis of Variance into More than Two Portions," he writes on the case where each observation belongs to one class of type A and to a different class of type B:

In such a case we can find separately the variance between classes of type A and between classes of type B; the balance of the total variance may represent only the variance within each subclass, or there may be in addition an interaction of causes, so that a change in class of type A does not have the same effect in all B classes. If the observations do not occur singly in the subclasses, the variance within the subclasses may be determined independently, and the presence or absence of interaction verified.

Fisher does not give a mathematical formulation of this model, which today in its simplest form is written as

$$y_{ij} = \mu + \alpha_i + \beta_j + \varepsilon_{ij}, \ \sum \alpha_i = \sum \beta_j = 0, \ i = 1, \ldots, r, \ j = 1, \ldots, c,$$

assuming that there is no interaction, and that ε_{ij} is normal $(0, \sigma^2)$. The effect of factor A is thus described by the parameters $\{\alpha_i\}$, of factor B by $\{\beta_j\}$, and the error by σ^2. Fisher does not explain that this is a special case of the general linear model obtained by setting the independent variable x_{ij} equal to 1 or 0 at the appropriate places, and that the estimates are obtained by the method of least squares. By means of examples, mainly randomized blocks and Latin squares for agricultural field trials, he shows how the total sum of squares may be partitioned. For the model above we get Table 20.3.

The effect of factor A is tested by means of $F = s_1^2/s_3^2$, and similarly for B.

Table 20.3. Analysis of variance for a two-way classification without replication.

Variance	Degr. Freedom	Sums of Squares	Mean Square
Rows (A)	$r - 1$	$c \sum_1^r (\overline{y}_{i.} - \overline{y}_{..})^2$	s_1^2
Columns (B)	$c - 1$	$r \sum_1^c (\overline{y}_{.j} - \overline{y}_{..})^2$	s_2^2
Error	$(r-1)(c-1)$	$\sum_1^r \sum_1^c (y_{ij} - \overline{y}_{i.} - \overline{y}_{.j} + \overline{y}_{..})^2$	s_3^2
Total	$rc - 1$	$\sum_1^r \sum_1^c (y_{ij} - \overline{y}_{..})^2$	

From then on, the analysis of variance became one of the most-used statistical techniques because of its compact form and easy interpretation in terms of causes of variation. The scope of this method was greatly enlarged by the publication of Fisher's [85] *The Design of Experiments* with its emphasis on the coordination of the design and analysis of factorial experiments.

There are no references to previous theory in *Statistical Methods* [73], which may be explained by the fact that the book was written for research workers who were supposed to accept the methods presented on the authority of Fisher. Nevertheless, it would have been helpful, not only for those readers but also for statisticians, if Fisher had offered some remarks on the connection between his and the older well-known methods. A large part of the book is based on the method of least squares, but this method is mentioned only casually in § 46.

In [72] the reader will find references only to Pearson [203] and "Student" [254]. The explanation may be that Fisher at the time was virtually ignorant of the literature before 1900. We make a few remarks on the history of the methods used by Fisher.

The partitioning of the total sum of squares into two components, representing the variation due to regression and error, respectively, goes back to Gauss and Laplace in their discussions on the methods of least squares. Thereafter it was used and refined by many others, particularly in connection with the fitting of orthogonal polynomials. The complete partitioning into sums of squares of orthogonal components was given by Thiele ([257], [260]) and Pizzetti [218], and rediscovered by Fisher. He did not elaborate this point for the different models considered. This gap in the mathematical foundation for the analysis of variance was filled out by Irwin ([125], [126]) and Cochran [32].

The Likelihood Function, Ancillarity, and Conditional Inference

21.1 The Amount of Information, 1925

Inspired by the large-sample results

$$V(\hat{\theta}|\theta) = 1/i(\theta) \text{ and } e_t = 1/\left[i(\theta)V(t|\theta)\right],$$

Fisher proposes to extend the concept of efficiency to finite samples. He defines "the amount of information" in the sample and in the statistic t as $i_x(\theta) = E[l'_x(\theta)]^2$ and $i_t(\theta) = E[l'_t(\theta)]^2$, respectively, and proves several properties that give this concept an intuitive appeal:

(1) If $f(x|\theta)$ is independent of θ then $i_x = 0$.

(2) If t is sufficient then i_t takes on the maximum value i_x.

(3) The information in an estimate can never exceed the information in the sample, that is, $i_t \leq i_x$.

(4) Independent observations supply amounts of information that are additive. He then defines the efficiency of t as $e_t = i_t/i_x$ for any sample size.

Fisher's idea was later used in the Cramér–Rao inequality, which says that, if $E(t) = \alpha(\theta)$, say, then $V(t|\theta) \geq [\alpha'(\theta)]^2/i(\theta)$.

When no sufficient statistics exist, some loss of information will necessarily result from using a single estimate of the parameter. Using the expansion

$$l'_x(\theta) = (\theta - \hat{\theta})l''_x(\hat{\theta}) + \cdots, \qquad (21.1)$$

Fisher proves that the loss of information by using $\hat{\theta}$ as estimate of θ equals

$$i_x - i_{\hat{\theta}} = V(\hat{\theta})V\left[l''_x(\theta)\,|l'_x(\theta)\right];$$

that is, the loss of information is proportional to $V(\hat{\theta})$ and the rate of change with respect to θ depends as indicated on the conditional variance of $l''_x(\theta)$. The new concepts are thus expressed in terms of the first and second derivatives of the likelihood function.

21.2 Ancillarity and Conditional Inference

In case a sufficient statistic does not exist, Fisher [74] remarks:

> Since the original data cannot be replaced by a single statistic, without
> loss of accuracy, it is of interest to see what can be done by calculat-
> ing, in addition to our estimate, an ancillary statistic which shall be
> available in combination with our estimate in future calculations.

Using one more term in the expansion (21.1.1) he notes that the variance
of $l'_x(\theta)$ decreases from the order of n to n^{-1} by conditioning on both $l'_x(\hat{\theta})$ and
$l''_x(\theta)$. He illustrates the use of ancillary statistics by estimating the location
parameter in the double exponential distribution with known scale parameter.
He begins with the ordered sample, $x_{(1)} \le x_{(2)} \le \cdots \le x_{(n)}$, which is sufficient, and introduces the ancillary statistics $a_i = x_{(i+1)} - x_{(i)}$, $i = 1, \ldots, n-1$,
which he calls the configuration of the sample. Because the distribution of a_i
is independent of θ, he gets

$$p(x|\theta) = p(\hat{\theta}, a|\theta)|J| = p(\hat{\theta}|a, \theta)h(a)|J|,$$

where $h(a)$ and the Jacobian J are independent of θ. Hence, all the information
on θ is contained in the conditional distribution of $\hat{\theta}$ for given a. This means
that $\hat{\theta}$ is sufficient for θ if we restrict the sample space to the points having
the same value of a as the given sample. The information can, however, vary
greatly from the one configuration to the other.

A more practical example of the usefulness of ancillary statistics is given in
Fisher [84], where he considers the conditional distribution of the contents of
a 2×2 contingency table for given marginal totals. This leads to the so-called
exact test for the odds ratio based on the hypergeometric distribution for the
only free variable in the table.

As another example he mentions the regression in a bivariate normal distribution, where the distribution of the regression coefficient is considered for
given values of the independent variable.

Although Fisher used conditional inference on many later occasions he did
not develop a general theory for this kind of statistical analysis.

21.3 The Exponential Family of Distributions, 1934

If t is sufficient then $p(x|\theta) = p(t|\theta)\alpha(x)$, say. Fisher uses this factorization of
the likelihood function to prove that, if a sufficient statistic exists, then the
density of x may be written as

$$f(x|\theta) = \exp\{a(\theta)b(x) + c(\theta) + d(x)\},$$

which is called the exponential family of distributions. This general form cov-
ers many of the common and most useful distributions, such as the normal, the
Gamma, the binomial, and the Poisson. The statistic $t = \sum b(x_i)$ is sufficient
for θ.

21.4 The Likelihood Function

Fisher's greatest achievement in statistical inference is the introduction of the likelihood function. It is implied by Laplace [148] that all the information in the observations regarding the unknown parameters in a statistical model is contained in the posterior distribution. Pointing out that Laplace's assumption of a uniform prior is superfluous, Fisher agreed, he renormed and renamed the posterior distribution as the likelihood function.

Fisher [82] continues the discussion of ancillary statistics and remarks that

> ... successive portions of the loss [of information] may be recovered by using as ancillary statistics, in addition to the maximum likelihood estimate, the second and higher differential coefficients at the maximum. In general we can only hope to recover the total loss, by taking into account the entire course of the likelihood function.

Jeffreys, the foremost advocate of inverse probability at the time, was greatly influenced by Fisher's use of invariance and likelihood. Commenting on the formula

$$p(\theta|x)d\theta = f(\theta)d\theta p(x|\theta),$$

Jeffreys [127] remarks:

> The whole of the information relevant to the unknowns and contained in the observations is expressed in the likelihood $[p(x|\theta)]$; hence if a set of statistics is such that the likelihood can be expressed in terms of it, it is equivalent to the likelihood; if it is not, there must be a sacrifice of information and loss of accuracy. This does not require that n should be large; it is true for all values of n. The method of moments sacrifices information because the moments to order 4 by themselves are not sufficient to make the likelihood calculable.

Another great achievement was Fisher's derivation of all the sampling distributions under normality and several other sampling distributions.

In the first instance he used the likelihood function only to find $\hat{\theta}$, the maximum likelihood estimate, for which he derived the asymptotically normal sampling distribution. Later, however, he realized the incompatibility of likelihood and repeated sampling from the same population. He then turned to inference based on the entire likelihood function, which he considered as indicating the strength of the support for the various values of the parameter. He ([86], p. 71) gives the binomial as a simple example:

> In the case under discussion a simple graph of the Mathematical Likelihood expressed as a percentage of its maximum, against the possible values of the parameter p, shows clearly enough what values of the parameter have likelihoods comparable with the maximum, and outside what limits the likelihood falls to levels at which the corresponding values of the parameter become implausible.

The "plausibility" of θ is thus tied up with the value of $l(\theta)$. He does not introduce the term "likelihood interval" but the quotation implies that a likelihood interval (θ_1, θ_2) with likelihood coefficient c may be defined as $\{\theta_1 < \theta < \theta_2 | l(\theta) > c\}$ in competition with the credibility and the confidence intervals.

A theory of inference based on the likelihood function has been developed by Edwards ([49], [51]).

Epilogue

The reason for stopping our account of the history of statistical inference about 1930 is the diversification that took place about that time. Before 1922 inference was based on either direct or inverse probability. To this was added inference based on the likelihood function by Fisher in 1922. A theory of testing statistical hypotheses was introduced by Neyman and E.S. Pearson ([188],[189]) and developed by Wald [264] into a general decision theory from a frequentist point of view. Inspired by Fisher's demand of invariance, Jeffreys [128] and Perks [214] attempted to find an "objective" prior distribution corresponding to the statistical model at hand. De Finetti [59] introduced prior distributions based on exchangeability, which inspired Savage [229] to develop a personalistic theory of decision. All these lines of thought are represented in the diagram at the end of Section 1.5.

Terminology and Notation

We have used the standard notation with $P(A)$ for the probability of the event A and $p(x)$ for the probability density of the frequency function for the random variable x. The symbol p is used generically so that we write $p(x, y) = p(x)p(y|x)$ for the bivariate density as the product of the marginal and the conditional density.

We use Latin letters for random variables and Greek letters for parameters. However, estimates of the parameter β, say, may be denoted by b, $\tilde{\beta}$, and $\hat{\beta}$ when comparing different estimates.

The expectation, variance, and covariance are written as $E(x), V(x)$, and $CV(x, y)$, respectively. To indicate that $E(x)$ depends on the parameter θ, we write $E(x|\theta)$; the same symbol is used for the conditional expectation when both x and θ are random variables.

Random errors are denoted by ε, and residuals by e.

The empirical and theoretical moments of order r are denoted by m_r and μ_r, respectively.

The inner product of the two vectors x and y may be written in one of the three forms $[xy], \sum x_i y_i, x'y$, the first being due to Gauss.

The standardized density of the normal distribution is denoted by ϕ and the distribution function by Φ. For the characteristic function we use ψ.

The most quoted book is Laplace's *Théorie analytique des probabilités* [159], abbreviated to TAP. *Oeuvres* and *Oeuvres completes* are indicated as O and OC, and collected papers as CP.

References

Books on the History of Statistics

Czuber, E.: Die Entwicklung der Wahrscheinlichkeitstheorie und ihrer Anwendungen. *Jahresber. Deutsch. Mat.-Ver., Vol. 7,* Teubner, Leipzig (1899). Reprinted by Johnson Reprint Corporation, New York (1960).

Dale, A.I.: *A History of Inverse Probability. From Thomas Bayes to Karl Pearson.* Springer, New York (1991).

David, H.A. and Edwards, A.W.F.: *Annotated Readings in the History of Statistics.* Springer, New York (2001).

Farebrother, R.W.: *Fitting Linear Relationships. A History of the Calculus of Observations, 1750–1990.* Springer, New York (1998).

Hald, A.: *A History of Probability and Statistics from 1750 to 1930.* Wiley, New York (1990).

Heyde, C.C. and Seneta, E.: *I.J. Bienaymé: Statistical Theory Anticipated.* Springer, New York (1977).

Kotz, S. and Johnson, N.L., Eds.: *Breakthroughs in Statistics. Vol. I. Foundations and Basic Theory.* Springer, New York (1992).

Kotz, S. and Johnson, N.L., Eds.: *Breakthroughs in Statistics. Vol. III.* Springer, New York (1997).

Lubbock, J.W. and Drinkwater-Bethune, J.E.: *On Probability.* Baldwin and Cradock, London (1830).

Pearson, K.: *The History of Statistics in the 17th and 18th Centuries.* Lectures of Karl Pearson given at University College London during the Academic Sessions 1921–1933. E.S. Pearson, ed., Griffin, London (1978).

Pizzetti, P.: I fondamenti matematici per la critica dei risultati sperimentali. *Atti Reg. Univ. Genova,* **11**, 113–333 (1892). Reprinted as Vol. 3 in *Biblioteca di Statistica* (1963).

Schneider, I.: *Die Entwicklung der Wahrscheinlichkeitstheorie von den Anfängen bis 1933. Einfürungen und Texte.* Wissenschaftliche Buchgesellschaft, Darmstadt (1988).

Sheynin, O.: *The History of the Theory of Errors.* Hänsel-Hohenhausen. Engelsbach (1996). (Deutsche Hochschulschriften, 1118).

Stigler, S.M.: *The History of Statistics. The Measurement of Uncertainty before 1900.* The Belknap Press of Harvard University Press, Cambridge, MA. (1986).

Todhunter, I.: *A History of the Mathematical Theory of Probability from the Time of Pascal to that of Laplace.* Macmillan, London (1865).

Books on the History of Statistical Ideas

Adams, W.J.: *The Life and Times of the Central Limit Theorem.* Kaedmon, New York (1974).

Cullen, M.J.: *The Statistical Movement in Early Victorian Britain.* Barnes & Noble, NewYork (1875).

Daston, L.J.: *Classical Probability in the Enlightenment.* Princeton Univ. Press, Princeton, New Jersey (1988).

Droesbeke, J.-J. et Tassi, P.: *Histoire de la Statistique.* Presses Univ. de France, Paris (1990).

Gigerenzer, G. et al.: *The Empire of Chance.* Camb. Univ. Press, Cambridge (1989).

Gouraud, C.: *Histoire du Calcul des Probabilitiés.* Durand, Paris (1848).

Heyde, C.C. and Seneta, E., Eds.: *Statisticians of the Centuries.* Springer, New York (2001).

Johnson, N.L. and Kotz, S.: *Leading Personalities in Statistical Sciences.* Wiley, New York (1997).

Kendall, M. and Plackett, R.L., Eds.: *Studies in the History of Statistics and Probability. Vol. II.* Griffin, London (1977).

Krüger, L., Daston, L.J. and Heidelberger, M., Eds.: *The Probabilistic Revolution. Vol. 1. Ideas in History.* MIT Press, Cambridge, MA, U.S.A. (1987).

Krüger, L., Gigerenzer, G. and Morgan, M.S., Eds.: *The Probabilistic Revolution. Vol. 2. Ideas in the Sciences.* MIT Press, Cambridge MA, U.S.A. (1987).

Mackenzie, D.A.: *Statistics in Britain. 1865–1930.* Edinb. Univ. Press, UK (1981).

Maistrov, L.E.: *Probability Theory. A Historical Sketch.* Academic Press, New York (1974).

Pearson, E.S. and Kendall, M., Eds.: *Studies in the History of Statistics and Probability. Vol. 1.* Griffin, London (1970).

Peters, W.S.: *Counting for Something. Statistical Principles and Personalities.* Springer, New York (1987).

Porter, T.M.: *The Rise of Statistical Thinking. 1820–1900.*Princeton Univ. Press, Princeton, NJ, U.S.A. (1986).

Stigler, S.M.: *Statistics on the Table. The History of Statistical Concepts and Methods.* Harvard Univ. Press, Cambridge, MA, U.S.A. (1999).

Walker, H.M.: *Studies in the History of Statistical Method.* Williams & Wilkins, Baltimore (1929).

Westergaard, H.: *Contributions to the History of Statistics.* King, London (1932).

General References

1. Adams, W.J.: *The Life and Times of the Central Limit Theorem*. Kaedmon, New York (1974).
2. Barton, D.E. and Dennis, K.E.: The conditions under which Gram–Charlier and Edgeworth curves are positive definite and unimodal. *Biometrika*, **39**, 425–427 (1952).
3. Bayes, T. : An essay towards solving a problem in the doctrine of chances. *Phil. Trans.*, 1763, **53**, 370–418 (1764). Reprinted in facsimile in *Two Papers by Bayes*, ed. W.E. Deming, (1940). Reprinted in *Biometrika*, **45**, 293–315 (1958); in Pearson and Kendall, eds.: (1970), and in Thomas and Peach, eds.: (1983). Translated into German with a commentary by H. E. Timerding (1908). Translated into French by J.P. Cléro (1988) with a preface by B. Bru, notes and postface by Cléro.
4. Bayes, T.: A demonstration of the second rule in the Essay towards the Solution of a Problem in the Doctrine of Chances. Published in the *Phil. Trans., Vol. LIII* (1765). Communicated by the Rev. Mr. Richard Price, in a letter to Mr. John Canton, M.A., F.R.S. *Phil. Trans.*, **54**, 296–325 (1974).
5. Bennett, J.H., ed.: *Natural Selection, Heredity, and Eugenics. Including Selected Correspondence of R. A. Fisher with Leonard Darwin and Others*. Clarendon Press, Oxford (1983).
6. Bernoulli, J.: *Ars Conjectandi*. Thurnisius, Basilea (1713). Reprinted in *Editions Culture et Civilisation*, Bruxelles (1968) and in *Die Werke von Jakob Bernoulli*, Band 3, Birkhäuser, Basel (1975). German translation by R. Haussner (1899). Part 1 translated into French by L. G. F. Vastel (1801) and into Italian by Dupont and Roero (1984). English translation of Part 2 by F. Maseres (1795) and of Part 4 by Bing Sung (1966). Russian translation of Part 4 by J.V. Uspensky (1913), reprinted in (1986).
7. Bernoulli, N.: *Letters to Montmort* (1710–1713); see Montmort (1713).
8. Bertrand, J.: *Calcul des Probabilités*. Gauthier-Villars, Paris (1889) 2nd ed. (1907). Reprinted by Chelsea, New York (1972).
9. Bessel, F.W.: *Abhandlungen von Friedrich Wilhelm Bessel*. 3 vols. Engelmann, Leipzig (1875–1876).
10. Bessel, F.W.: *Fundamenta astronomiae pro anno MDCCLV*. Regiomonti. (1818).
11. Bessel, F.W.: Untersuchungen über die Wahrscheinlichkeit der Beobachtungsfehler. *Astron. Nachrichten*, **15**, No. 385–359, 369–404 (1838). Reprinted in *Abhandlungen*, **2**.
12. Bienaymé, I.J.: Mémoire sur la probabilité des résultats moyens des observations; démonstration directe de la règle de Laplace. *Mém. Acad. Roy. Sci. Inst. France*, **5**, 513–558 (1838).
13. Bienaymé, I.J.: Sur la probabilité des erreurs d'après la methode des moindres carrés. *Liouville's J. Math. Pures Appl.*, (1), **17**, 33–78 (1852).
14. Bienaymé, I.J.: Considérations á l'appui de la découverte de Laplace sur la loi de probabilité dans la méthode des moindres carrés. *C.R. Acad. Sci., Paris,* **37**, 309–324 (1853), and in Liouville's *J. Math. Pures Appl.*, (2), **12**, 158–176 (1867).

15. Bing, F.: Om aposteriorisk Sandsynlighed. (On posterior probability.) *Tidsskrift for Mathematik*, 4th Series, **3**, 1–22, 66–70, 122–131 (1879).

16. Boole, G.: *The Laws of Thought*. Macmillan, London (1854). Reprinted by Dover, New York (1958).

17. Boscovich, R.J.: De Litteraria Expeditione per Pontificiam Ditionem, et Synopsis Amplioris Operis. *Bononiensi Scientiarum et Artum Instituto atque Academia Commentarii*, **4**, 353–396 (1757).

18. Boscovich, R.J. and Maire, C.: *De Litteraria Expeditione per Pontificiam Ditionem ad Dimetiendas duas Meridiani Gradus*. Palladis, Rome (1755). French translation in Boscovich and Maire (1770).

19. Boscovich, R.J. and Maire, C.: *Voyage astronomique et géographique dans l'état de l'église*. Tilliard, Paris (1770).

20. Bowditch, N.: *Mécanique céleste,* translation into English of Laplace's Traité, Vols. 1–4, with commentaries (1829–1839). Reprinted by Chelsea, New York, as *Celestial Mechanics* (1966).

21. Box, J.F.: R.A. Fisher: *The Life of a Scientist*. Wiley, New York (1978).

22. Bravais, A.: Analyse mathématique sur les probabilitiés des erreurs de situation d'un point. *Mém. Acad. Roy. Sci. Inst. France*, **9**, 255–332 (1846).

23. Cam, L., Le: The central limit theorem around 1935. (with discussion). *Statist. Sci.*, **1**, 78–96 (1986).

24. Charlier, C.V.L.: Über die Darstellung willkürlicher Funktionen. *Ark. Mat. Astr. Fys.*, 2, No. **20**, 1–35 (1905).

25. Charlier, C.V.L.: *Grunddragen av den matematiska statistiken*. Lund, Sverige (1910).

26. Charlier, C.V.L.: *Die Grundzüge der Mathematischen Statistik*. Lund, Sverige (1920).

27. Chauvenet, W.: *On the Method of Least Squares. An Appendix to A Manual of Spherical and Practical Astronomy*, Vol. **2**, 469–566. Lippincott, Philadelphia (1863). Issued separately (1868).

28. Chebyshev, P.L.: *Oeuvres de P.L. Tchebychef*, ed. by A. Markov et N. Sonin. 2 vols. (1899–1907). French translation of Russian edition. Reprinted by Chelsea, New York. References to the Russian journals are from *Oeuvres*.

29. Chebyshev, P.L.: Démonstration élémentaire d'une proposition génerale de la théorie des probabilités. *Crelle's J. reine und angew. Math.*, **33**, 259–267 (1846). Oeuvres, **1**, 17–26.

30. Chebyshev, P.L.: (1855). Sur les fractions continues [in Russian]. *J. Math. Pure et Appliquées,* **3**, 289–323 (1858). *Oeuvres,* **1**, 203–230.

31. Chebyshev, P.L.: Des valeurs moyennes. *Liouville's J. Math. Pures et Appl.*, (2) **12**, 177–184 (1867). *Oeuvres,* **1**, 687–694.

32. Cochran, W.G.: The distribution of quadratic forms in a normal system with applications to the analysis of covariance. *Proc. Camb. Phil. Soc.*, **30**, 178–191 (1934).

33. Cournot, A.A.: *Exposition de la Théorie des Chances et des Probabilités*. Hachette, Paris (1843). Reprinted in *Oeuvres Complètes, Tome 1*, B. Bru ed., (1984), Librairie J. Vrin, Paris.

34. Crosland, M.: *The Society of Arcueil*. Heinemann, London (1967).

35. Cullen, M.J.: *The Statistical Movement in Early Victorian Britain*. Barnes & Noble, New York (1975).

36. Czuber, E.: *Theorie der Beobachtungsfehler*. Teubner, Leipzig (1891).
37. Czuber, E.: Die Entwicklung der Wahrscheinlichkeitstheorie und ihrer Anwendungen. Jahresber. *Deutsch. Mat.-Ver.*, Vol. **7**, Teubner, Leipzig (1899). Reprinted by Johnson Reprint Corporation, New York (1960).
38. Dale, A.I.: *A History of Inverse Probability. From Thomas Bayes to Karl Pearson*. Springer, New York (1991).
39. Daston, L.: *Classical Probability in the Enlightenment*. Princeton University Press, Princeton, New Jersey (1988).
40. David, H.A.: First (?) occurrence of common terms in mathematical statistics. *Amer. Statistician*, **49**, 121–133 (1995).
41. David, H.A. and Edwards, A.W.F.: *Annotated Readings in the History of Statistics*. Springer, New York (2001).
42. Droesbeke, J.-J. et Tassi, P.: *Histoire de la Statistique*. Presses Univ. de France, Paris (1990)
43. Edgeworth, F.Y.: The method of least squares. *Phil. Mag., Ser. 5*, **16**, 360–375 (1883).
44. Edgeworth, F.Y.: Correlated averages. *Phil. Mag., Ser. 5*, **34**, 190–204 (1892).
45. Edgeworth, F.Y.: Note on the calculation of correlation between organs. *Phil. Mag., Ser. 5*, **36**, 350–351 (1893).
46. Edgeworth, F.Y.: The law of error. *Trans. Camb. Phil. Soc.*, **20**, 35–65 and 113–141 (1905).
47. Edgeworth, F.Y.: On the probable error of frequency constants. *J. Roy. Statist. Soc.*, **71**, 381–397, 499–12, 651–68 (1908).
48. Edgeworth, F.Y.: Addendum on probable errors on frequency constants. *J. Roy. Statist. Soc.*, **72**, 81–790 (1909).
49. Edwards, A.W.F.: *Likelihood*. Camb. Univ. Press, Cambridge (1972).
50. Edwards, A.W.F.: The history of likelihood. *Intern. Statist. Rev.*, **42**, 9–15 (1974). Reprinted in *Likelihood* (1992).
51. Edwards, A.W.F.: *Likelihood*. Expanded Edition. Johns Hopkins Univ. Press, Baltimore (1992).
52. Edwards, A.W.F.: Three early papers on efficient parametric estimation. *Statist. Sci.*, **12**, 35–47 (1997).
53. Eisenhart, C.: Karl Pearson. In: *Dictionary of Scientific Biography*, ed. C.C. Gillispie, Vol. **10**, 447–473 (1974).
54. Elderton, W.P.: *Frequency-Curves and Correlation*. Layton, London (1906). 2nd ed. (1927). 3rd ed. by Camb. Univ. Press, Cambridge (1938).
55. Ellis, R.L.: On the foundations of the theory of probabilities. *Trans. Camb. Phil. Soc.*, **8**, 1–6 (1849).
56. Encke, J.F.: Über die Methode der kleinsten Quadrate. *Berliner Astron. Jahrbuch für* 1834, 249–312; für 1835, 253–320; für 1836, 253–308 (1832–1834).
57. Engledow, F.L and Yule, G.U.: The determination of the best value of the coupling ratio from a given set of data. *Proc. Camb. Phil. Soc.* **17**, 436–440 (1914).
58. Farebrother, R.W.: *Fitting Linear Relationships. A History of the Calculus of Observations*. 1750–1900. Springer, New York (1998).
59. Finetti, B. de: La prévision: ses lois logiques, ses sources subjectives. *Ann. Institut Henri Poincaré*, **7**, 1–68 (1937).

60. Fisher, R.A.: *Collected Papers of R. A. Fisher.* Ed. by J.H. Bennett. Univ. Adelaide, Australia. 5 Volumes (1971–1974). Referred to as CP plus the number of the paper.

61. Fisher, R.A.: *Mendelism and Biometry* (1911). Manuscript published in J.H. Bennett 51–58 (1983).

62. Fisher, R.A.: On an absolute criterion for fitting frequency curves. *Messenger Math.*, **41**, 155–160 (1912). CP 1. Reprinted in *Statist. Sci.*, **12**, 39–41 (1997).

63. Fisher, R.A.: Frequency distribution of the values of the correlation coefficient in samples from an indefinitely large population. *Biometrika*, **10**, 507–521 (1915). CP 4.

64. Fisher, R.A.: The correlation between relatives on the supposition of Mendelian inheritance. *Trans. Roy. Soc. Edinb.*, **52**, 399–433 (1918). CP 9.

65. Fisher, R.A.: A mathematical examination of the methods of determining the accuracy of an observation by the mean error, and by the mean square error. *Monthly Notices Roy. Astron. Soc.*, **80**, 758–770 (1920). CP 12.

66. Fisher, R.A.: On the "probable error" of a coefficient of correlation deduced from a small sample. *Metron*, **1**, 3–32 (1921). CP 14.

67. Fisher, R.A.: On the mathematical foundations of theoretical statistics. *Phil. Trans.*, A, **222**, 309–368 (1922). CP 18.

68. Fisher, R.A.: On the interpretation of χ^2 from contingency tables, and the calculation of P. *J. Roy. Statist. Soc.*, **85**, 87–94 (1922). CP 19.

69. Fisher, R.A.: The goodness of fit of regression formulæ, and the distribution of regression coefficients. *J. Roy. Statist. Soc.*, **85**, 597–612 (1922). CP 20.

70. Fisher, R.A.: The distribution of the partial correlation coefficient. *Metron*, **3**, 329-332 (1924). CP 35.

71. Fisher, R.A.: On a distribution yielding the error functions of several well known statistics. *Proc. Intern. Congress Math., Toronto*, **2**, 805–813 (1924). CP 36. (Published 1928).

72. Fisher, R.A.: The influence of rainfall on the yield of wheat at Rothamsted. *Phil. Trans.*, B. **213**, 89–142 (1924). CP 37.

73. Fisher, R.A.: *Statistical Methods for Research Workers.* Oliver and Boyd, Edinburgh (1925). Later editions (1928), (1930), (1932), (1934), (1936), (1938), (1941), (1944), (1946), (1950), (1954), (1958), (1970). The fourteenth edition (1970), is reprinted as part of *Statistical Methods, Experimental Design and Scientific Inference,* Oxford Univ. Press, Oxford (1990).

74. Fisher, R.A.: Theory of statistical estimation *Proc. Camb. Phil. Soc.*, **22**, 700–725 (1925). CP 42.

75. Fisher, R.A.: Applications of "Student's" distribution. *Metron*, **5**, No. 3, 90–104 (1925). CP 43.

76. Fisher, R.A.: Sur la solution de l'équation intégrale de M.V. Romanovsky. *C.R. Acad. Sci. Paris*, **181**, 88–89 (1925). CP 46.

77. Fisher, R.A.: The arrangement of field experiments. *J. Min. Agric. G. B.*, **33**, 503–513 (1926). CP 48.

78. Fisher, R.A.: On a distribution yielding the error functions of several well known statistics. *Proc. Intern. Congress Math., Toronto*, **2**, 805–813 (1928). CP 36. Presented to the Intern. Congress Math. in 1924.

79. Fisher, R.A.: The general sampling distribution of the multiple correlation coefficient. *Proc. Roy. Soc. London*, A, **121**, 654–673 (1928). CP 61.

80. Fisher, R.A.: Inverse probability. *Proc. Camb. Phil. Soc.*, **26**, 528–535 (1930). CP 84.

81. Fisher, R.A.: *The Genetical Theory of Natural Selection*. Oxford Univ. Press, Oxford (1930). 2nd ed. by Dover, New York (1958).

82. Fisher, R.A.: Two new properties of mathematical likelihood. *Proc. Roy. Soc. London*, A, **144**, 285–307 (1934). CP 108.

83. Fisher, R.A.: The mathematical distributions used in the common tests of significance. *Econometrica*, **3**, 353–365 (1935). CP 123.

84. Fisher, R.A.: The logic of inductive inference (with discussion). *J. Roy. Statist. Soc.*, **98**, 39–82 (1935). CP 124.

85. Fisher, R.A.: *The Design of Experiments*. Oliver and Boyd, Edinburgh (1935). Later editions (1937), (1942), (1947), (1949), (1951), (1960), (1966). The eight edition, (1966), is reprinted as part of *Statistical Methods, Experimental Design and Scientific Inference*, Oxford Univ. Press, Oxford (1990).

86. Fisher, R.A.: *Statistical Method and Scientific Inference*. Oliver and Boyd, Edinburgh (1956). 2nd ed. (1959), 3rd ed. (1973). The third edition is reprinted as part of *Statistical Methods, Experimental Design and Scientific Inference*, Oxford Univ. Press, Oxford (1990).

87. Fisher, R.A.: *Statistical Methods, Inference, and Experimental Design*. Oxford Univ. Press (1990). Reprints of the latest editions of [73], [85], and [86].

88. Fisher, R.A. and Yates, F.: *Statistical Tables for Biological, Agricultural and Medical Research*. Oliver and Boyd, Edinburgh (1938). Later editions (1943), (1948), (1953), (1957), (1963).

89. Forest, E.L. de: On an unsymmetrical probability curve. *Analyst, 9*, 135–142, 161–168; **10**, 1–7, 67–74 (1882–1883). Reprinted in Stigler [248]

90. Forest, E.L. de: On an unsymmetrical law of error in the position of a point in space. *Trans. Connecticut Acad. Art and Sciences, 6*, 123–138 (1884). Reprinted in Stigler [248].

91. Forest, E.L. de: On the law of error in target shooting. *Trans. Connecticut Acad. Art and Sciences, 7*, 1–8 (1885).

92. Galton, F.: Regression towards mediocrity in hereditary stature. *J. Anthrop. Inst.*, **15**, 246–263 (1886).

93. Galton, F.: Family likeness in stature. *Proc. Roy. Soc. London*, **40**, 42–73 (1886). Appendix by J.D. Hamilton Dickson, 63–66.

94. Galton, F.: *Natural Inheritance*. Macmillan, London (1889). Reprinted by AMS Press, New York, (1973).

95. Galton, F.: Co-relations and their measurement, chiefly from anthropometric data. *Proc. Roy. Soc. London*, **45**, 135–145 (1889).

96. Galton, F.: Kinship and correlation. *North Amer. Rev.*, **150**, 419–431 (1890). Reprinted in *Statist. Sci.*, **4**, 81–86 (1989).

97. Galton, F.: A geometric determination of the median value of a system of normal variants, from two of its centiles. *Nature*, **61**, 102–104 (1899).

98. Gauss, C.F.: *Werke*. 12 vols. 1863–1933. Königliche Gesellschaft der Wissenschaften zu Göttingen. Reprinted by Olms, Hildesheim (1973).

99. Gauss, C.F.: *Disquisitiones Arithmeticae, Werke* **1** Königliche Gesellschaft der Wissenschaften zu Göttingen, (1870).

100. Gauss, C.F.: *Theoria motus corporum coelestium in sectionibus conicis solem ambientium.* Perthes et Besser, Hamburg. *Werke,* **7**, 1–280 (1809). Translated by C.H. Davis as *Theory of the Motion of the Heavenly Bodies Moving about the Sun in Conic Sections,* Little, Brown and Co., Boston, 1857. Reprinted by Dover, New York, 1963. Translated into German by C. Haase, Hannover (1865).

101. Gauss, C.F.: Disquisito de elementis ellipticis palladis. *Comment. Recent. Soc. Scient. Göttingen,* **1**, 26 pp. *Werke,* **6**, 3–24 (1810).

102. Gauss, C.F.: Bestimmung der Genauigkeit der Boebachtungen. *Z. Astron. und verwandte Wiss.,* **1**, 185–216 (1816). *Werke,* **4**, 109–117.

103. Gauss, C.F.: Theoria combinationis observationum erroribus minimis obnoxiae. Pars prior, et Pars posterior. *Comm. Soc. Reg. Gottingensis Rec.,* **5**, 33–62, 63–90. Read (1821) and (1823). *Werke,* **4**, 3–26, 29–53. Reports on Pars prior in Göttingische gelehrte Anzeigen, 1821, and on Pars posterior in 1823, *Werke,* **4**, 95–100, and 100–104. Translated by G.W. Stewart with an introduction and afterword as *Theory of the Combination of Observations Least Subject to Errors.* In: *Classics in Applied Mathematics,* SIAM, Philadelphia.

104. Gauss, C.F.: Supplementum theoriae combinationis observationum erroribus minimis obnoxiae. *Comm. Soc. Reg. Gottingensis Rec.,* **6**, 57–93 (1828). Read 1826. *Werke,* **4**, 57–93. Report in Göttingische gelehrte Anzeigen (1826), *Werke,* **4**, 104–108.

105. Gauss, C.F.: Letter to Bessel, 28 February 1839. *Briefwechsel zwischen Gauss und Bessel,* 523–525. Engelmann, Leipzig (1880) and *Gauss Werke,* **8**, 146–147.

106. Gauss, C.F.: Letter to Schumacher, 25 November 1844. *Gauss, Werke,* **8**, 147–148.

107. Gigerenzer, G. et al.: *The Empire of Chance.* Camb. Univ. Press, Cambridge (1989).

108. Gosset, W.S., see Student.

109. Gouraud, C.: *Histoire du calcul des probabilités.* Durand, Paris (1848).

110. Gram, J.P.: *Om Rækkeudviklinger, bestemte ved Hjælp af de mindste Kvadraters Methode.* Høst, Kjøbenhavn (1879).

111. Gram, J.P.: *Über die Entwickelung reeller Functionen in Reihen mittelst der Methode der kleinsten Quadrate. J. reine angew. Math.,* **94**, 41–73 (1883).

112. Hagen, G.H.L.: *Grundzüge der Wahrscheinlichkeits-Rechnung.* Dümler, Berlin (1837).

113. Hald, A.: *A History of Probability and Statistics and Their Applications before* 1750. Wiley, New York (1990).

114. Hald, A.: *A History of Mathematical Statistics From* 1750 *to* 1930. Wiley, New York (1998).

115. Hald, A.: The early history of the cumulants and the Gram–Charlier series. *Intern. Statist. Rev.,* **68**, 137–153 (2000).

116. Hald, A.: On the history of the correction for grouping. *Scand. J. Statistics,* **28**, 417–428 (2001).

117. Hald, A.: On the history of series expansions of frequency functions and sampling distributions, 1873-1944. Matematisk-Fysiske Meddelelser. *The Roy. Danish Acad. Sci. Lett.* 88 pp. C.A. Reitzels Forlag, Copenhagen (2002).

118. Hartley, D.: *Observations on Man, His Frame, His Duty, and His Expectations.* Richardson, London (1749). Reprinted by Scholar's Fascimiles and Reprints, Gainesville, Florida (1966) with an introduction by T.L. Huguelet.

119. Helmert, F.R.: *Die Ausgleichungsrechnung nach der Methode der kleinsten Quadrate* (1872). 2nd ed. (1907), 3rd ed. (1924). Teubner, Leipzig.

120. Helmert, F.R.: Über die Berechnung des wahrscheinlichen Fehlers aus einer endlichen Anzahl wahrer Beobachtungsfehler. *Z. f. Math. und Physik,* **20**, 300–303 (1875).

121. Helmert, F.R.: Über die Wahrscheinlichkeit der Potenzsummen der Beobachtungsfehler und über einige damit im Zusammenhange stehende Fragen. *Z. Math. und Physik,* **21**, 192–218 (1876).

122. Helmert, F.R.: Die Genauigkeit der Formel von Peters zur Berechnung des wahrscheinlichen Beobachtungsfehler direchter Beobachtungen gleicher Genauigkeit. *Astron. Nachr.,* **88**, 113–132 (1876).

123. Heyde, C.C. and Seneta, E.: I.J. *Bienaymé: Statistical Theory Anticipated.* Springer, New York (1977).

124. Heyde, C.C. and Seneta, E., Eds.: *Statisticians of the Centuries.* Springer, New York (2001).

125. Irwin, J.O.: Mathematical theorems involved in the analysis of variance. *J. Roy. Statist. Soc.,* **94**, 284–300 (1931).

126. Irwin, J.O.: On the independence of the constituent items in the analysis of variance. *J. Roy. Statist. Soc. Suppl.,* **1**, 236–251 (1934).

127. Jeffreys, H.: Maximum likelihood, inverse probability and the method of moments. *Ann. Eugenics,* **8**, 146–151 (1938).

128. Jeffreys, H.: An invariant form for the prior probability in estimation problems. *Proc. Roy. Soc.,* A, **186**, 453–461 (1946). Reprinted in *Collected Papers,* 6, (1977).

129. Jevons, W.S.: *The Principles of Science.* London (1877). 1st ed. (1874).

130. Johnson, N.L. and Kotz, S.: *Leading Personalities in Statistical Sciences.* Wiley, New York (1997).

131. Jordan, C.: Sur la probabilité des épreuves répétées. Le théoreme de Bernoulli et son inversion. *Bull. Soc. Math. France,* **54**, 101–137 (1926).

132. Kapteyn, J.C.: *Skew Frequency Curves in Biology and Statistics.* Noordhoff, Groningen (1903).

133. Kapteyn, J.C. and Uven, M.J. van: *Skew Frequency Curves in Biology and Statistics.* Hoitsema Brothers, Groningen (1916).

134. Karlin, S.: R.A. Fisher and evolutionary theory. *Statist. Sci.,* 7, 13–33 (1992).

135. Kendall, M.G. and Stuart, A.: *The Advanced Theory of Statistics.* Griffin, London (1958).

136. Kendall, M. and Plackett, R.L., Eds.: *Studies in the History of Statistics and Probability.* Vol. II. Griffin, London (1977).

137. Keynes, J.M.: *A Treatise on Probability.* Macmillan, London (1921). Reprinted (1951), (1952). Reprinted as Vol. VIII of *The Collected Writings of John Maynard Keynes* (1973).

138. Khintchine, A.Ya.: Sur la loi des grand nombres. *C.R. Acad. Sci. Paris,* 477–479 (1929).

139. Kotz, S. and Johnson, N.L., eds.: *Breakthroughs in Statistics.* Vol. I. Foundations and Basic Theory. Springer, New York (1992).

140. Kotz, S. and Johnson, N.L., Eds.: *Breakthroughs in Statistics. Vol. II. Methodology and Distribution.* Springer, New York (1992).

141. Kotz, S. and Johnson, N.L., Eds.: *Breakthroughs in Statistics. Vol. III.* Springer, New York (1997).

142. Kries, J. von: *Die Principien der Wahrscheinlichkeitsrechnung.* Freiburg. Reprinted 1927 by Mohr, Tübingen (1886).

143. Kruskal, W.: The significance of Fisher: A review of *R.A. Fisher: The Life of a Scientist. J. Amer. Statist. Assoc.,* **75**, 1019–1030 (1980).

144. Krüger, L., Daston, L.J. and Heidelberger, M., Eds.: *The Probabilistic Revolution. Vol. 1. Ideas in History.* MIT Press, Cambridge, MA, U.S.A. (1987).

145. Krüger, L., Gigerenzer, G. and Morgan, M.S., Eds.: *The Probabilistic Revolution.* Vol. **2**. *Ideas in the Sciences.* MIT Press, Cambridge, MA, U.S.A. (1987).

146. Lagrange, J.L.: Mémoire sur l'utilité de la méthode de prendre le milieu entre les résultats de plusieurs observations. *Misc. Taurinensia,* **5**, 1770–1773, 167–232 (1776). *Oeuvres,* **2**, 173–234, Paris.

147. Laplace, P.S.: *Oeuvres complètes.* 14 vols. 1878–1912. Gauthier-Villars, Paris. Page references to the works of Laplace are to this edition cited as OC.

148. Laplace, P.S.: Mémoire sur la probabilité des causes par les événements. *Mém. Acad. Roy. Sci. Paris* (Savants étrangers), **6**, 621–656 (1774). OC **8**, 27–65. Translated into English with an introduction by S.M. Stigler in *Statistical Science,* **1**, 359–378 (1986).

149. Laplace, P.S.: Mémoire sur les probabilités. *Mém. Acad. Roy. Sci. Paris,* 1778, 227–332 (1781), OC **9**, 383–485.

150. Laplace, P.S.: Mémoire sur les approximations des formules qui sont fonctions de très grand nombres. *Mém. Acad. Roy. Sci. Paris,* 1782, 1–88 (1785), OC **10**, 209–291.

151. Laplace, P.S.: Mémoire sur les approximations des formules qui sont fonctions de très grand nombres. (Suite). *Mém. Acad. Roy. Sci. Paris,* 1783, 423–467 (1786), OC **10**, 295–338.

152. Laplace, P.S.: Théorie de Jupiter et de Saturne. *Mém. Acad. Roy. Sci. Paris,* 1785, 33–160 (1788), OC **11**, 95–239.

153. Laplace, P.S.: Sur quelques points du système du monde. *Mém. Acad. Roy. Sci. Paris,* 1789, 1–87 (1793), OC **11**, 477–558.

154. Laplace, P.S.: *Traité de mécanique céleste* 4 vols. Paris (1799–1805). OC **1–4**. Translated into English by N. Bowditch, Boston (1829–1839). Reprinted by Chelsea, New York (1966).

155. Laplace, P.S.: Mémoire sur les approximations des formules qui sont fonctions de très grands nombres et sur leur application aux probabilités. *Mém. Acad. Sci. Paris,* 1809, 353–415 (1810). OC **12**, 301–345.

156. Laplace, P.S.: Supplément au Mémoire sur les approximations des formules qui sont fonctions de trés grands nombres. *Mém. Acad. Sci. Paris,* 1809, 559–565 (1810). OC **12**, 349–353.

157. Laplace, P.S.: Mémoire sur les intégrales définies et leur application aux probabilités, et spécialement a la recherche du milieu qu'il faut choisir entre les résultats des observations. *Mém. Acad. Sci. Paris,* 1810, 279–347 (1811). OC **12**, 357–412.

158. Laplace, P.S.: Du milieu qu'il faut choisir entre les résultats d'un grand nombre d'observations. *Conn. des Temps,* 213–223 (1811). OC **13**, 78.

159. Laplace, P.S.: *Théorie analytique des probabilités.* Courcier, Paris (1812). 2nd ed. (1814). 3rd ed. (1820). Quatriéme Supplement (1825). OC **7**.

160. Laplace, P.S.: *Essai philosophique sur les probabilités.* Paris (1814). Sixth edition translated by Truscott, F.W. and F.L. Emory as *A Philosophical Essay on Probabilities* (1902). Reprinted (1951) by Dover, New York. Fifth edition (1825) reprinted with notes by B. Bru (1986), Bourgois, Paris. Fifth edition translated by A.I. Dale as *Philosophical Essays on Probabilities,* with notes by the translator (1995), Springer, New York.

161. Laplace, P.S.: Sur l'application du calcul des probabilités à la philosophie naturelle. *Premier Supplément,* TAP, OC **7**, 497–530 (1816).

162. Laplace, P.S.: Application du calcul des probabilités aux opérations géodésiques. *Deuxieme Supplément,* TAP, OC **7**, 531–580 (1818).

163. Laplace, P. S.: *Traité de mécanique céleste.* Vol. **5**, Paris (1825). OC **5**. Reprinted by Chelsea, New York (1969).

164. Lauritzen, S.L.: *Thiele. Pioneer in Statistics.* Oxford Univ. Press, Oxford (2002).

165. Legendre, A.M.: *Nouvelles méthodes pour la determination des orbites des comètes.* Courcier, Paris (1805). Reissued with supplements in 1806 and 1820. Four pages from the appendix on the method of least squares translated into English in *A Source Book in Mathematics,* ed. D.E. Smith, pp. 576–579, McGraw-Hill, New York; reprinted by Dover, New York (1959).

166. Lindeberg, J.W.: Eine neue Herleitung des Exponentialgesetzes in der Wahrscheinlichkeitsrechnung. *Math. Zeit.,* **15**, 211–225 (1922).

167. Lipps, G.F.: Die Theorie der Collectivgegenstände. *Wundt's Philos. Studien,* **17**, 79–184, 467–575 (1901).

168. Lipps, G.F.: *Die Theorie der Collectivgegenstände.* Engelmann, Leipzig (1902). Reprint of Lipps. [167]

169. Lubbock, J.W.: On the calculation of annuities, and on some questions in the theory of chances. *Trans. Camb. Phil. Soc.,* **3**, 141–155 (1830). Reprinted in the *J. Inst. Actuaries,* **5**, 197–207 (1855).

170. Lubbock, J.W. and Drinkwater-Bethune, J.E.: *On Probability. Baldwin and Cradock,* London (1830).

171. Lüroth, J.: Vergleichung von zwei Werten des wahrscheinlichen Fehlers. *Astron. Nachr.,* **87**, 209–220 (1876).

172. MacKenzie, D.A.: *Statistics in Britain.* 1865–1930. Edinb. Univ. Press, UK (1981).

173. Maistrov, L.E.: *Probability Theory. A Historical Sketch.* Academic Press, New York (1974).

174. Mayer, T.: Abhandlung über die Umwälzung des Monds um seine Axe und die scheinbare Bewegung der Mondsflecten. *Kosmographische Nachrichten und Sammlungen auf das Jahr* 1748, **1**, 52–183 (1750).

175. Merriman, M.: A list of writings relating to the method of least squares, with historical and critical notes. *Trans. Connecticut Acad. Art and Sciences,* **4**, 151–232 (1877). Reprinted in Stigler (1980).

176. Merriman, M.: *A Text-Book on the Method of Least Squares.* Wiley, New York (1884). References are to the 8th edition, (1915).

177. Mises, R. von: Fundamentalsätze der Wahrscheinlichkeitsrechnung. *Math. Z.,* **4**, 1–97 (1919).

178. Mises, R. von: *Wahrscheinlichkeitsrechnung und ihre Anwendung in der Statistik und theoretischen Physik.* Deuticke, Leipzig (1931). Reprinted by Rosenberg, New York (1945).

179. Mises, R. von: *Mathematical Theory of Probability and Statistics.* Edited and complemented by H. Geiringer. Academic Press, New York (1964).

180. Moivre, A. de: *The Doctrine of Chances: or, A Method of Calculating the Probability of Events in Play.* Pearson, London (1718).

181. Moivre, A. de: *Annuities upon Lives: or, The Valuation of Annuities upon any Number of Lives; as also, of Reversions,* to which is added, *An Appendix concerning the Expectations of Life, and Probabilities of Survivorship.* Fayram, Motte and Pearson, London (1725).

182. Moivre, A. de: *Miscellanea Analytica de Seriebus et Quadraturis.* Tonson & Watts, London (1730). Miscellaneis Analyticis Supplementum.

183. Moivre, A. de: *Approximatio ad Summam Terminorum Binomii* $(a+b)^n$ *in Seriem expansi* (1733). Printed for private circulation.

184. Moivre, A. de: *The Doctrine of Chances.* The second edition, fuller, clearer, and more correct than the first. Woodfall, London (1738). Reprinted by Cass, London (1967).

185. Moivre, A. de: *The Doctrine of Chances.* The third edition, fuller, clearer, and more correct than the former. Millar, London (1756). Reprinted by Chelsea, New York (1967).

186. Morant, G.M.: *A Bibliography of the Statistical and Other Writings of Karl Pearson. Biometrika* Office, London (1939).

187. Morgan, A. de: *An Essay on Probabilities and on their application to Life Contingencies and Insurance Offices.* Longman, Orme, Brown, Green & Longmans, London (1838). Reprinted by Arno Press, New York (1981).

188. Neyman, J. and Pearson, E.S.: On the use and interpretation of certain test criteria for purposes of statistical inference. Part I. *Biometrika,* **20A**, 175–240 (1928).

189. Neyman, J. and Pearson, E.S.: On the use and interpretation of certain test criteria for purposes of statistical inference. Part II. *Biometrika,* **20A**, 263–294 (1928).

190. Neyman, J. and Scott, E.L.: Consistent estimates based on partially consistent observations. *Econometrica,* **16**, 1–32 (1948).

191. Pearson, E.S.: *Karl Pearson: An Appreciation of Some Aspects of His Life and Work.* Camb. Univ. Press, Cambridge (1938).

192. Pearson, E.S.: Some incidents in the early history of biometry and statistics 1890-94. *Biometrika,* **52** 3–18 (1965). Reprinted in Pearson and Kendall (1970).

193. Pearson, E.S.: Some reflexions on continuity in the development of mathematical statistics, 1885–1920. *Biometrika,* **54**, 341–355 (1967). Reprinted in Pearson and Kendall (1970).

194. Pearson, E.S.: Some early correspondence between W.S. Gosset, R.A. Fisher and Karl Pearson, with notes and comments. *Biometrika,* **55**, 445–457 (1968). Reprinted in Pearson and Kendall (1970).

195. Pearson, E.S. and Hartley, H.O.: *Biometrika Tables for Statisticians. Volume 1.* Camb. Univ. Press, Cambridge (1954).

196. Pearson, E.S. and Hartley, H.O. eds.: *Biometrika Tables for Statisticians. Volume 2.* Camb. Univ. Press, Cambridge (1972).

197. Pearson, E.S. and Kendall, M. eds.: *Studies in the History of Statisticians and Probability*, **1**. Griffin, London (1970).

198. Pearson, K.: *Karl Pearson's Early Statistical Papers*. Ed. E.S. Pearson. Camb. Univ. Press, Cambridge (1948).

199. Pearson, K.: *The Grammar of Science*. W. Scott, London (1892). 2nd ed. (1900), 3rd ed. (1911).

200. Pearson, K.: Contribution to the mathematical theory of evolution. *Phil. Trans., A,* **185**, 71–110 (1894). Reprinted in Karl Pearson's Early Statistical Papers, 1–40. Camb. Univ. Press, Cambridge (1948).

201. Pearson, K.: Contributions to the mathematical theory of evolution. II. Skew variations in homogeneous material. *Phil. Trans.,* A, **186**, 343–414 (1895). Reprinted in *Karl Pearson's Early Statistical Papers*, 41–112. *Camb. Univ. Press, Cambridge* (1948).

202. Pearson, K.: Mathematical contributions to the theory of evolution. III. Regression, heredity and panmixia. *Phil. Trans., A,* **187**, 253–318 (1896). Reprinted in *Karl Pearson's Early Statistical Papers*, Camb. Univ. Press, Cambridge (1948).

203. Pearson, K.: On the criterion that a given system of deviations from the probable in the case of a correlated system of variables is such that it can be reasonably supposed to have arisen from random sampling. *Phil. Mag.,* (5), **50**, 157–175 (1900). Reprinted in *Karl Pearson's Early Statistical Papers*, Camb. Univ. Press, Cambridge (1948).

204. Pearson, K.: On the probable errors of frequency constants. *Biometrika,* **2**, 273–281 (1903).

205. Pearson, K.: *Mathematical contributions to the theory of evolution. XIII. On the theory of contingency and its relation to association and normal correlation*. Drapers' Company Research Memoirs. Biometric Series, I (1904). Reprinted in *Karl Pearson's Early Statistical Papers*, Camb. Univ. Press, Cambridge (1948).

206. Pearson, K.: *Mathematical contributions to the theory of evolution. XIV. On the general theory of skew correlation and non-linear regression*. Drapers' Company Research Memoirs. Biometric Series, II (1905). Reprinted in Karl Pearson's Early Statistical Papers, Camb. Univ. Press, Cambridge (1948).

207. Pearson, K., ed.: *Tables for Statisticians and Biometricians*. Camb. Univ. Press, Cambridge (1914). 2nd ed. (1924), 3rd ed. (1930).

208. Pearson, K.:. *The Life, Letters and Labours of Francis Galton. 4 Parts: I (1914), II (1924), III A (1930), III B* (1930) Camb. Univ. Press, Cambridge

209. Pearson, K., ed.: *Tables of the Incomplete Γ-Function*. His Majesty's Stationery Office, London (1922). Reprinted (1934).

210. Pearson, K., ed.: *Tables for Statisticians and Biometricians. Part II.* Camb. Univ. Press, Cambridge (1931)

211. Pearson, K.: *Tables of the Incomplete Beta-Function*. Prepared under the direction of and edited by K. Pearson. Camb. Univ. Press, Cambridge (1934). Second edition with a new introduction by E. S. Pearson and N. L. Johnson (1968).

212. Pearson, K.: *The History of Statistics in the 17th and 18th Centuries*. Lectures by Karl Pearson given at University College London during the

academic sessions 1921–1933. Edited by E. S. Pearson. Griffin, London (1978).

213. Pearson, K. and Filon, L.N.G.: Mathematical contributions to the theory of evolution, IV. On the probable errors of frequency constants and on the influence of random selection on variation and correlation. *Phil. Trans.*, A, **191**, 229–311 (1898). Reprinted in *Karl Pearson's Early Statistical Papers*, 179–261, Camb. Univ. Press, Cambridge (1948).

214. Perks, W.: Some observations on inverse probability, including a new indifference rule. *J. Inst. Actuaries*, **73**, 285–312. Discussion, 313–334. (1947).

215. Peters, C.A.F.: Über die Bestimmung des wahrscheinlichen Fehlers einer Beobachtung aus den Abweichungen der Beobachtungen von ihrem arithmetischen Mittel. *Astron. Nachr.*, **44**, 29–32 (1856).

216. Peters, W.S.: *Counting for Something. Statistical Principles and Personalities.* Springer, New York (1987).

217. Pfanzagl, J. and Sheynin, O.: Studies in the history of probability and statistics XLIV. A forerunner of the *t*-distribution. *Biometrika,* **83**, 891–898 (1996).

218. Pizzetti, P.: I fundamenti matematici per la critica dei risultati sperimentali. Atti della Regia Università de Genova, **11**, 113–333 (1892). Reprinted as Vol. **3** in *Biblioteca di Statistica* (1963). Page references are to the reprint.

219. Plackett, R.L.: The discovery of the method of least squares. *Biometrika,* **59**, 239–251 (1972). Reprinted in Kendall and Plackett, (1977).

220. Poisson, S.D.: Sur la probabilité des résultats moyens des observations. *Conn. des Tems pour* 1827, 273–302 (1824).

221. Poisson, S.D.: Suite du Mémoire sur la probabilité du résultat moyen des observations, inséré dans la Connaissance des Tems de l'année 1827. *Conn. des Tems pour* 1832, 3–22 (1829).

222. Poisson, S.D.: *Recherches sur la Probabilité des Jugements en Matière Criminelle et en Matière Civile, précedées des Règles Générales du Calcul des Probabilités.* Bachelier, Paris (1837). Translated into German by C.H. Schnuse as *Lehrbuch der Wahrscheinlichkeitsrechnung und deren wichtigsten Anwendungen*, Braunschweig (1841).

223. Porter, T.M.: *The Rise of Statistical Thinking.* 1820–1900. Princeton Univ. Press, Princeton, NJ, U.S.A. (1986).

224. Prevost, P. and Lhuilier, S.A.J.: Sur les probabilités. *Classe Math. Mém. Acad. Roy. Sci. et Belles-Lettres,* 1796, Berlin, 117–142 (1799a).

225. Prevost, P. and Lhuilier, S.A.J.: Mémoire sur l'art d'estimer la probabilité des causes par les effects. *Classe Phil. Spéculative Mém. Acad. Roy. Sci. et Belles-Lettres,* 1796, Berlin, 3–24 (1799b).

226. Price, R.: Introduction and appendix to Bayes' Essay. *Phil. Trans,* 1763, **53**, 370–375 and 404–418 (1764).

227. Price, R.: A demonstration of the second rule in Bayes' Essay. *Phil. Trans.*, 1764, **54**, 296–297 and 310–325 (1765).

228. Rao, C.R.: R.A. Fisher: The founder of modern statistics. *Statist. Sci.*, **7**, 34-48 (1992).

229. Savage, L.J.: *The Foundations of Statistics.* Wiley, New York (1954).

230. Savage, L.J.: On rereading R.A. Fisher (with discussion). Edited by J.W. Pratt. *Ann. Statist.*, **4**, 441–500 (1976). Reprinted in *The Writings of L.J. Savage*, Amer. Statist. Assoc. and Inst. Math. Statist. (1981).

231. Schneider, I.: *Die Entwicklung der Wahrscheinlichkeitstheorie von den Anfängen bis 1933. Einfürungen and Texte.* Wissenschaftliche Buchgellschaft, Darmstadt (1988).

232. Schols, C.M.: Over de theorie der fouten in de ruimte en in het platte vlak. *Verh. Kon. Akad. Wetens.*, Amsterdam, **15**, 67 pp. French version (1886): Théorie des erreurs dans le plan et dans l'espace. *Ann. École Polytech. Delft*, **2**, 123–178 (1875) .

233. Schols, C.M.: La loi de l'erreur résultante. *Ann. École Polytech. Delft*, **3**, 140–150 (1887).

234. Sheppard, W.F.: On the calculation of the most probable values of frequency constants, for data arranged according to equidistant divisions of a scale. *Proc. London Math. Soc.*, **29**, 353–380 (1898).

235. Sheppard, W.F.: On the application of the theory of error to cases of normal distribution and normal correlation. *Phil. Trans.*, A, **192**, 101–167 (1899).

236. Sheynin, O.: *The History of the Theory of Errors.* Hänsel-Hohenhausen. Engelsbach (1996). (Deutsche Hochschulschriften, 1118.)

237. Simpson, T.: *Miscellaneous Tracts on Some Curious, and Very Interesting Subjects in Mechanics, Physical-Astronomy, and Speculative Mathematics.* Nourse, London (1757).

238. Smith, K.: On the best values of the constants in frequency distributions. *Biometrika*, **11**, 262–276 (1916).

239. Snedecor, G.W.: *Calculation and Interpretation of Analysis of Variance and Covariance.* Collegiate Press, Ames, Iowa (1934).

240. Soper, H.E.: On the probable error of the correlation coefficient to a second approximation. *Biometrika*, **9**, 91–115 (1913).

241. Soper, H.E., A.W. Young, B.M. Cave, A. Lee, and K. Pearson: On the distribution of the correlation coefficient in small samples. Appendix II to the papers of "Student" and R.A. Fisher. A cooperative study. *Biometrika*, **11**, 328–413 (1917).

242. Sprott, D.A.: Gauss's contributions to statistics. *Historia Mathematica*, **5**, 183–203 (1978).

243. Steffensen, J.F.: *Matematisk Iagttagelseslære.* Gad, København (1923).

244. Steffensen, J.F.: *Interpolationslære.* Gad, København (1925). English ed. Interpolation. Williams & Wilkin, Baltimore (1927).

245. Steffensen, J.F.: The theoretical foundation of various types of frequency-functions. The English School; Karl Pearson's types. The Continental School; the A-series; Charlier's B-series. Some notes on factorial moments, pp. 35-48 in *Some Recent Researches in the Theory of Statistics and Actuarial Science*, Camb. Univ. Press, Cambridge (1930).

246. Steffensen, J.F.: *Forsikringsmatematik.* Gad, København (1934).

247. Stigler, S.M.: Laplace, Fisher, and the discovery of the concept of sufficiency. *Biometrika*, **60**, 439–445 (1973). Reprinted in Kendall and Plackett (1977).

248. Stigler, S.M. ed.: *American Contributions to Mathematical Statistics in the Nineteenth Century.* 2 Vols. Arno Press, New York (1980).

249. Stigler, S.M.: Thomas Bayes's Bayesian inference. *J. Roy. Statist. Soc. Ser.* A, **143**, 250–258 (1982).

250. Stigler, S.M.: The History of Statistics: *The Measurement of Uncertainty before* 1900. The Belknap Press of Harvard University Press, Cambridge, MA (1986).

251. Stigler, S.M.: Laplace's 1774 memoir on inverse probability. *Statist. Sci.*, **1**, 359–378 (1986).

252. Stigler, S.M.: *Statistics on the Table. The History of Statistical Concepts and Methods.* Harvard Univ. Press, Cambridge, Massachussetts (1999).

253. Stirling, J.: *Methodus differentialis.* London (1730).

254. Student: The probable error of a mean. *Biometrika,* **6**, 1–25 (1908). Reprinted in *"Student's" Collected Papers.*

255. Student: Probable error of a correlation coefficient. *Biometrika,* **6**, 302–310 (1908). Reprinted in *"Student's" Collected Papers.*

256. *"Student's" Collected Papers. Ed. by E. S. Pearson and J. Wishart.* Issued by the *Biometrika* Office, University College, London. Univ. Press, Cambridge (1942).

257. Thiele, T.N.: *Almindelige Iagttagelseslære: Sandsynlighedsregning og mindste Kvadraters Methode. (The General Theory of Observations: Probability Calculus and the Method of Least Squares.)* Reitzel, København (1889); see Lauritzen (2002).

258. Thiele, T.N.: *Elementær Iagttagelseslære.* Gyldendal, København (1897).

259. Thiele, T.N.: Om Iagttagelseslærens Halvinvarianter. (On the halfinvariants in the theory of observations.) *Kgl. danske Videnskabernes Selskabs Forhandlinger*, 1899, Nr. **3**, 135–141 (1899).

260. Thiele, T.N.: *Theory of Observations.* Layton, London (1903). Reprinted in *Ann. Math. Statist.* **2**, 165–307 (1931).

261. Todhunter, I.: *A History of the Mathematical Theory of Probability from the Time of Pascal to that of Laplace.* London, Macmillan (1865).

262. Uspensky, J.V.: *Introduction to Mathematical Probability.* McGraw-Hill, New York (1937).

263. Venn, J.: *The Logic of Chance.* London (1866). 2nd ed. (1876), 3rd ed. (1888). Reprinted by Chelsea, New York (1962).

264. Wald, A.: *Statistical Decision Functions.* Wiley, New York (1950).

265. Walker, H.M.: *Studies in the History of Statistical Method.* Williams & Wilkins, Baltimore (1929).

266. Westergaard, H.: *Die Grundzüge der Theorie der Statistik.* Fischer, Jena (1890).

267. Westergaard, H.: *Contributions to the History of Statistics.* King, London (1932).

268. Wishart, J.: The generalized product moment distribution in samples from a normal multivariate population. *Biometrika,* **20A**, 32–52, 424 (1928).

269. Yates, F. and Mather, K.: Ronald Aylmer Fisher, 1890–1962. *Biographical Memoirs of Fellows of the Royal Society of London,* **9**, 91–120 (1963). Reprinted in *Collected Papers of R.A. Fisher,* **1**, 23 52.

270. Yule, G.U.: On the significance of Bravais' formulæ for regression, etc., in the case of skew correlation. *Proc. Roy. Soc. London,* **60**, 477–489 (1897a).

271. Yule, G.U.: On the theory of correlation. *J. Roy. Statist. Soc.,* **60**, 812–854 (1897b).

272. Yule, G.U.: An investigation into the causes of changes in pauperism in England, chiefly during the last two intercensal decades. I. *J. Roy. Statist. Soc.,* **62**, 249–295 (1899).

273. Yule, G.U.: On the theory of correlation for any number of variables, treated by a new system of notation. *Proc. Roy. Soc.*, A, **79**, 182–193 (1907).

274. Yule, G.U.: *An Introduction to the Theory of Statistics.* Griffin, London (1911). 11th edition with M.G. Kendall as coauthor in (1937).

275. Zabell, S.L.: The rule of succession. *Erkenntnis*, **31**, 283–321 (1989a).

276. Zabell, S.L.: R.A. Fisher on the history of inverse probability. *Statist. Sci.*, **4**, 247–263 (1989b).

Subject Index

Absolute criterion for fitting frequency curves, 163

Absolute deviation, 2, 3, 42, 49

Absolute moments, 64, 66, 172

Analysis of variance, 96, 154, 161, 162, 171, 177, 185, 188–192

Analytic probability theory, 34

Ancillary statistic, 7, 42, 177, 194, 195

Antrophometric measurements, 122, 134–136, 138, 139, 141, 145

Arc length of meridian, 50, 52

Asymptotic expansion of densities and integrals, 3, 18, 20, 29, 34, 37, 38, 44, 46

Asymptotic normality
of linear combinations, 5
of posterior distributions, 34, 39, 44
of sampling distributions, 38, 67, 97, 179

Averages, method of, 2, 48

Bayes's postulate, 28

Bayes's rule, 25, 27, 168

Bayes's theorem, 36, 74, 76, 163

Best linear asymptotically normal estimate, 86, 88, 89

Beta distribution, 27, 38, 44

Beta probability integral, 28

Beta-binomial distribution, 40

Binomial distribution, 3, 11, 12, 14, 17–20, 22, 187, 194

Bivariate normal distribution, 46, 85, 126, 131, 133–136, 138, 139, 141, 166, 170, 183, 194

Cauchy distribution, 85, 155, 180

Central limit theorem, 4, 5, 15, 34, 56, 60, 61, 64, 65, 69, 83–86, 90–92, 99, 107, 115, 128, 149, 161, 176, 179

Characteristic functions, 34, 64, 84, 85, 89, 90, 149

Chi-squared, 7, 117
and likelihood function, 180
distribution of sample variance, 132, 133, 152–154, 160, 161, 185–188
exponent of multivariate normal, 123, 132
test for goodness of fit, 118, 122–125, 164

Coin tossings, data and theory, 74, 77

Conditional distribution and sufficiency, 6, 90, 172, 173, 175–177

Conditional inference, 7, 193, 194

Confidence ellipsoids, 132

Confidence intervals, 6, 22, 23, 63, 64, 77, 87, 94, 97, 132, 133, 154, 196

Configuration of sample, 172, 194

Consistency of estimate, 6, 38, 176, 177

Continuity correction, 24, 92

Contour ellipse, 136

Convolution formula, 83, 99, 149

Correlation coefficient
bivariate, 7, 80, 126, 131, 139, 141–144, 165–169, 171
multiple, 145, 169–171, 189
partial, 7, 170

Correlation ratio, 190
Correlation, concept and generation of, 131, 134, 135, 139, 142, 143, 145, 146, 165, 173
Covariance, 85, 99, 124–126, 131
Cramér–Rao inequality, 193
Credibility limits, 39, 63–66, 71, 77, 95
Cumulant generating function, 90
Cumulants, 91, 111, 112, 115, 156

Decomposition
    of sums of squares, 98, 142, 153
    of variances, 188, 189, 191
Degrees of freedom, 70, 98, 124, 125, 132, 133, 152–154, 160, 172, 189, 190
Design of experiments, 161, 162, 177, 191
Differential equation
    for normal density, 57, 107, 111, 120
    for Pearson's distributions, 121, 123
Diffusion model, 90
Direct probability, 1, 97, 99, 105, 107, 155, 179
Dirichlet distribution, 67
Double exponential distribution, 2, 4, 42, 173

Efficiency of estimate, 6, 65, 88–90, 100, 123, 126, 131, 152, 160, 172, 176, 177, 180–182, 193
Elementary errors, 106, 107, 112
Ellipticity of the Earth, 50, 52
Empirical distributions, 58, 98, 135, 136, 138
Equations of condition, 47
Error distributions, 58, 61, 99, 100, 112
Estimation theory, direct probability
    Averages, method of, 2
    Largest absolute deviation, 2, 48
    Least absolute deviation, 2, 50, 52
    Least squares, 52, 53, 55, 56, 58, 60, 64, 85–88, 93–95, 97, 99–101, 105, 109, 112–115, 144, 146, 147, 150, 163, 171, 186, 189–192
    Linear unbiased minimum variance, 97, 98

Maximum likelihood, 106, 109, 159, 163, 165, 168, 175, 177, 178, 182, 183, 195
Minimum Chi-squared, 146
Selected points, method of, 47, 49
Exponential family of distributions, 194

F (variance ratio), 7, 98, 161, 162, 187–191
Factorization criterion, 176
Fechner distribution, 111
Fictitious observations, 37
Fiducial limits, 7, 23, 160

Games of chance, 11, 12, 77
Gamma distribution, 121, 133, 194
Generating function, 34, 90, 112
Geodetic applications, 52, 149
Geometric method of proof, 165, 167, 172
Goodness of fit, 24, 48, 118, 122–125, 180, 190
Gram–Charlier expansion, 92, 113–115
Gram–Schmidt orthogonalization, 146, 147
Graphical methods, 134–136, 138, 139

Helmert distribution, 151
Helmert's transformation, 151, 153
Heredity, 118, 119, 134, 136, 143
Hermite polynomials, 91
Hypergeometric distribution, 194
Hypothesis of elementary errors, 106, 108, 112
Hypothetical infinite population, 175

Incomplete beta function, 119
Incomplete gamma function, 119
Indifference principle, 73, 77, 79
Induction and probability, 27, 29, 35, 74
Information, 68, 71, 72, 74, 75, 90, 173, 176, 177, 181, 193–195
Interaction, 162, 177, 191
Interquartile range, 135
Intrinsic accuracy, 176
Inverse probability, 1–7, 24–26, 33–38, 42, 46, 56–59, 63, 65, 67, 69, 70, 72–80, 83, 85, 88, 97, 99–101, 105–108, 118, 155, 164, 165, 168, 178, 195

Inverse probability limits, 63

Kurtosis, 111

Largest absolute deviations, 2, 48
Latin square, 161, 191
Law of large numbers, 14, 15, 17, 25, 84
Least absolute deviation, 2, 48, 50, 52
Least squares, invention and justifica-
    tion, 5, 48, 50, 52, 53, 55, 56, 58,
    60, 61, 64, 86–89, 93–95, 97, 99,
    100, 105, 106, 108, 109, 112–114,
    163
Level of significance, 177
Likelihood function, 3, 7, 24, 37, 46, 79,
    107, 108, 177, 182, 193–196
Linear constraints, estimation under,
    96, 97
Linear minimum variance estimate, 73,
    83, 85, 93, 95, 99, 106
Location parameter, 1, 4, 42, 46, 56–60,
    86, 87, 99–101, 135, 180, 194
Location-scale family, estimation of
    parameters, 180
Log-likelihood function, 177
Loss functions, 3–7
Loss of information, 193

Maximum likelihood, 109
Maximum likelihood estimation, 45, 59,
    65, 72, 80, 108, 109, 159, 164, 168,
    175, 178–183, 195
Mean absolute difference, 152
Mean deviation, 17, 149, 172
Mean square error, 93, 152, 188, 190,
    191
Measurement error model, 3, 47, 52, 64,
    105
Median, 42, 75, 89, 99–101, 135, 136,
    139, 141, 180
Midparent, 135
Midrange, 181
Minimax method, 48
Minimum variance estimation, 5, 59,
    60, 73, 85–88, 93–95, 97–100, 106,
    176, 179
Moment generating function, 90, 112
Moments, method of, 111, 122, 123,
    160, 164, 180, 195

Multinomial distribution, 3, 22, 44, 67,
    99, 124, 125, 134, 180
Multiple correlation coefficient, 145,
    189, 190
Multivariate normal density, 23, 132
Multivariate normal distribution, 4, 22,
    23, 72, 85, 97, 123, 131–133, 141,
    142

Neyman–Pearson school, 160
Noncentral Chi-squared distribution,
    172
Normal deviate, 135
Normal distribution, 3, 4, 20, 56, 83–85,
    87, 131, 132, 135, 136, 142–144
Normal equations, 53, 59, 60, 87, 93, 94,
    96, 114, 145, 146
Normal probability integral, 135, 154
Normal probability paper, 135
Nuisance parameters, 42
Null hypothesis, 162, 177

Orthogonal
    functions, 113, 114
    polynomials, 115
    regression, 146, 147
    transformation, 132, 153, 185
Orthonormal decomposition, 153

Parametric statistical model, 1, 36, 37,
    41–43, 163, 175, 195
Pearson's family of distributions, 6, 117,
    120, 121, 123
Percentiles, 126, 135
Peter's formula, 152
Pivotal quantity, 7
Planning of observations, 162
Poisson distribution, 188
Posterior
    consistency, 38
    density, 7, 36, 57, 80, 97, 105, 164
    distribution, 3, 5, 25, 34, 35, 38, 42,
        44, 46, 59, 67, 68, 71, 72, 79, 80,
        97, 99–101, 108, 143, 168, 179, 195
    expected loss, 3
    mean, 4, 101
    median, 3, 42, 57, 60, 99
    mode, 4, 37, 42, 58, 59, 61, 65, 67, 68,
        71, 72, 86, 101, 168
    probability, 26, 74, 79, 97, 106

Prediction, 3, 34, 40, 41, 75, 78
Principle of inverse probability, 2, 4, 35, 36, 38, 40, 42, 57–60, 65, 67, 73, 100, 101, 106, 178
Prior distribution, 37, 40, 46, 56, 58, 65, 74, 79, 97, 149
Probable error, 118, 134, 138, 143, 169, 178

Quartile, 135, 141
Quincunx, 134

Randomization, 161, 162, 177
Randomized block experiment, 161, 162, 191
Rectangular distribution, 2, 42, 58, 83
Reduced normal equations, 59
Reduction of data, 175
Regression, 131–136, 138–140, 190
Replication, 161
Residuals, 98, 151, 170
Rule of succession, 41, 74, 75, 78

Sampling distributions under normality, 65, 72, 106, 107, 109, 149–151, 153, 154, 156, 171
Selected points, method of, 48, 49
Semicircular distribution, 42

Significance test, 2, 3, 161, 162, 169, 187, 189
Standard deviation, 15, 65, 66, 107, 118, 127, 143, 154, 155, 165, 167, 169, 172, 176, 177, 181, 185, 187
Standard meter, 52, 53
Standardized variable, 20, 115, 131, 139, 141
Statistic, 166, 171, 176, 177
Stature data, 139, 142
Studentization, 185
Sufficiency, 6, 90, 172, 173, 175–177
Sunrise, probability of, 28, 74

t distribution, 7, 69–71, 154–156, 162, 185, 186
Tail probability, 3, 12–15, 39
Terminology, 196
Transformation to normality, 6
Triangular distribution, 42
Triangulation, 56

Updating linear estimates, 95
Updating the prior distribution, 40

Variance ratio, 98, 187
Variance, estimate of, 149–152, 172, 182, 183
Variance, Fisher's definition of, 178

Author Index

Adams, W.J., 92, 201

Barton, D.E., 117, 201
Bayes, T., 1, 26–28, 35, 36, 201, 212, 213
Bennett, J.H., 160, 201, 204
Bernoulli, D., 105
Bernoulli, J., 11, 17–19, 38, 39, 77, 84, 201, 207
Bernoulli, N., 14, 201
Berthollet, C.L., 34
Bertillon, A., 138
Bertrand, J., 133, 201
Bessel, F.W., 58, 97, 98, 101, 112, 201, 206
Bienaymé, I.J., 15, 18, 67, 90, 91, 132, 201, 207
Bing Sung, 201
Bing, F., 79, 202
Bonaparte, N., 33
Boole, G., 78, 202
Boscovich, R.J., 2, 50, 51, 88, 202
Bowditch, N., 52, 202, 208
Box, J.F., 160, 202
Bravais, A., 131–133, 143, 202, 214
Bru, B., 201, 209
Buffon, G.L.L., 74

Cam, L. Le, 92, 202
Canton, J., 201
Cave, B. M., 213
Cave, B.M., 168
Charlier, C.V.L., 92, 111, 113, 114, 202
Chauvenet, W., 108, 202

Chebyshev, P.L., 14, 15, 18, 117, 146, 202
Cléro, J.P., 201
Cochran, W.G., 192, 202
Cournot, A.A., 36, 46, 73, 76, 78, 79, 202
Crosland, M., 34, 202
Cullen, M.J., 202
Czuber, E., 156, 203

Dale, A.I., 80, 203, 209
Darwin, C., 118, 159
Darwin, L., 159
Daston, L.J., 208
David, H.A., 177, 203
Davis, C.H., 206
Deming, W.E., 201
Dennis, K.E., 117, 201
Dickson, J.D.H., 136, 142, 205
Drinkwater-Bethune, J.E., 209
Droesbeke, J.-J., 203

Edgeworth, F.Y., 46, 68–72, 80, 92, 111, 142, 143, 155, 177, 179, 180, 203
Edwards, A.W.F., 105, 146, 196, 203
Eisenhart, C., 119, 203
Elderton, W.P., 121, 203
Ellis, R.L., 77, 78, 203
Emory, F.L., 209
Encke, J.F., 106, 149, 203
Engledow, F.L., 146, 203

Farebrother, R.W., 48, 203
Fechner, G.T., 115, 118

Filon, L.N.G., 68, 72, 118, 143, 165, 212
Finetti, B. de, 196, 203
Fisher, R.A., 1, 3, 23, 37, 68, 80, 90, 98, 106, 119, 120, 123–125, 155, 159, 160, 162–169, 172, 173, 175, 192–195, 202, 204, 205, 208, 210, 212–214
Forest, E.L. de, 133, 205

Galton, F., 111, 118, 119, 126, 134, 141, 142, 159, 165, 205, 211
Gauss, C.F., 1, 4, 5, 7, 55–60, 64, 65, 73, 85, 87, 88, 90, 93–97, 99, 101, 105, 108, 149, 152, 163, 172, 196, 205, 213
Gigerenzer, G., 206, 208
Gillispie, C.C., 203
Gosset, W.S., 70, 154–156, 160, 165, 185, 206, 210
Gouraud, C., 206
Gram, J.P., 92, 111–114, 118, 146, 206
Guiness, R.E., 159

Haase, C., 206
Hagen, G.H.L., 106, 108, 111, 112, 120, 164, 206
Hald, A., 29, 34, 44, 46, 80, 90, 92, 113, 122, 206
Hartley, D., 25, 26, 207
Hartley, H.O., 121, 210
Haussner, R., 201
Heidelberger, M., 208
Helmert, F.R., 133, 149–152, 156, 172, 185, 207
Heyde, C.C., 67, 207
Huguelet, T.L., 207
Hume, D., 74

Irwin, J.O., 192, 207

Jeffreys, H., 195, 196, 207
Jevons, W.S., 78, 207
Johnson, N.L., 207, 211
Jordan, C., 207
Jordan, Ch., 116

Kapteyn, J.C., 6, 111, 126, 207
Karlin, S., 160, 207
Kendall, M., 201, 207, 210–213

Kendall, M.G., 124, 207, 215
Keynes, J.M., 80, 207
Khintchine, A.Ya., 15, 207
Kotz, S., 207
Krüger, L., 208
Kries, J. von, 79, 208
Kruskal, W., 160, 208

Lüroth, J., 69, 155, 209
Lagrange, J.L., 22, 34, 67, 208
Laplace, P.S., 1–5, 7, 17, 24, 33–44, 46, 49, 51, 60, 61, 73, 74, 77–79, 83–88, 90, 93, 94, 97, 99, 101, 108, 131, 149, 172, 176, 178, 180, 195, 197, 201, 202, 208, 209, 213, 214
Lauritzen, S.L., 114, 209, 214
Lavoisier, A.L., 34
Lee, A., 168, 213
Legendre, A.M., 52, 55, 56, 209
Lexis, W., 111
Lhuilier, S.A.J., 76, 212
Lindeberg, J.W., 90, 209
Liouville, J., 201
Lipps, G.F., 113–116, 209
Lubbock, J.W., 78, 209

MacKenzie, D.A., 119, 209
Maire, C., 50, 202
Maistrov, L.E., 209
Markov, A., 202
Maseres, F., 201
Mather, K., 160, 214
Mayer, T., 2, 48, 49, 209
Median, 51
Merriman, M., 108, 109, 209
Mises, R. von, 67, 209
Moivre, A. de, 17–20, 26, 39, 84, 210
Montmort, P.R. de, 201
Morant, G.M., 119, 210
Morgan, A. de, 24, 26, 73, 210
Morgan, M.S., 208

Neyman, J., 160, 183, 196, 210

Pascal, B., 214
Pearson, E.S., 119, 121, 160, 196, 210, 211, 214
Pearson, K., 6, 24, 68, 72, 111, 117, 118, 122, 123, 126, 140, 142, 143, 159,

160, 163, 165, 168, 177, 189, 192, 203, 210, 211, 213
Perks, W., 196, 212
Peters, C.A.F., 152, 212
Peters, W.S., 212
Pfanzagl, J., 69, 212
Pizzetti, P., 153, 192, 212
Plackett, R.L., 56, 207, 212, 213
Poisson, S.D., 14, 15, 73, 75, 76, 78, 83, 85, 90, 91, 180, 212
Porter, T.M., 212
Pratt, J.W., 160, 212
Prevost, P., 76, 212
Price, R., 26, 28, 74, 78, 201, 212

Quetelet, A., 111, 126

Rao, C.R., 160, 212

Savage, L.J., 160, 196, 212
Schneider, I., 213
Schnuse, C.H., 212
Schols, C.M., 133, 213
Schumacher, H.C., 97, 206
Scott, E.L., 183, 210
Seneta, E., 67, 207
Sheppard, W.F., 122–125, 165, 213
Sheynin, O., 69, 212, 213
Simpson, T., 42, 213
Smith, D.E., 209
Smith, K., 146, 213
Snedecor, G.W., 187, 213
Sonin, N., 202
Soper, H.E., 143, 165, 166, 168, 213
Sprott, D.A., 87, 213

Steffensen, J.F., 113, 114, 116, 123, 213
Steward, G.W., 206
Stigler, S.M., 28, 34, 35, 43, 48, 50, 69, 90, 112, 205, 208, 209, 213, 214
Stirling, J., 18, 214

Sources and Studies in the
History of Mathematics and Physical Sciences

Continued from page ii

A.W. Grootendorst
Jan de Witt's *Elementa Curvarum Linearum, Liber Primus*

A. Hald
A History of Parametric Statistical Inference from Bernoulli to Fischer 1713–1935

T. Hawkins
Emergence of the Theory of Lie Groups: An Essay in the History of Mathematics 1869–1926

A. Hermann/K. von Meyenn/V.F. Weisskopf (Eds.)
Wolfgang Pauli: Scientific Correspondence I: 1919–1929

C.C. Heyde/E. Seneta
I.J. Bienaymé: Statistical Theory Anticipated

J.P. Hogendijk
Ibn Al-Haytham's *Completion of the Conics*

J. Høyrup
Length, Widths, Surfaces: A Portrait of Old Babylonian Alegbra and Its Kin

A. Jones (Ed.)
Pappus of Alexandria, Book 7 of the *Collection*

E. Kheirandish
The Arabic Version of Euclid's *Optics,* Volumes I and II

J. Lützen
Joseph Liouville 1809–1882: Master of Pure and Applied Mathematics

J. Lützen
The Prehistory of the Theory of Distributions

G.H. Moore
Zermelo's Axiom of Choice

O. Neugebauer
A History of Ancient Mathematical Astronomy

O. Neugebauer
Astronomical Cuneiform Texts

F.J. Ragep
Naṣīr al-Dīn al-Ṭūsī's *Memoir on Astronomy*
(al-Tadhkira fī ᶜilm al-hay'a)

B.A. Rosenfeld
A History of Non-Euclidean Geometry

G. Schubring
Conflicts Between Generalization, Rigor and Intuition: Number Concepts Underlying the Development of Analysis in 17th-19th Century France and Germany

Sources and Studies in the
History of Mathematics and Physical Sciences

Continued from the previous page

J. Sesiano
Books IV to VII of Diophantus' *Arithmetica*: In the Arabic Translation Attributed to Quṣṭā ibn Lūqā

L.E. Sigler
Fibonacci's *Liber Abaci*: A Translation into Modern English of Leonardo Pisano's Book of Calculation

J.A. Stedall
The Arithmetic of Infinitesimals: John Wallis 1656

B. Stephenson
Kepler's Physical Astronomy

N.M. Swerdlow/O. Neugebauer
Mathematical Astronomy in Copernicus's De Revolutionibus

G.J. Toomer (Ed.)
Appolonius *Conics* Books V to VII: The Arabic Translation of the Lost Greek Original in the Version of the Banū Mūsā, Edited, with English Translation and Commentary by G.J. Toomer

G.J. Toomer (Ed.)
Diocles on Burning Mirrors: The Arabic Translation of the Lost Greek Original, Edited, with English Translation and Commentary by G.J. Toomer

C. Truesdell
The Tragicomical History of Thermodynamics, 1822–1854

K. von Meyenn/A. Hermann/V.F. Weisskopf (Eds.)
Wolfgang Pauli: Scientific Correspondence II: 1930–1939

K. von Meyenn (Ed.)
Wolfgang Pauli: Scientific Correspondence III: 1940–1949

K. von Meyenn (Ed.)
Wolfgang Pauli: Scientific Correspondence IV, Part I: 1950–1952

K. von Meyenn (Ed.)
Wolfgang Pauli: Scientific Correspondence IV, Part II: 1953–1954

Printed in the United States of America